MORE UFOs
OVER WARMINSTER

MORE UFOs
OVER
WARMINSTER

ARTHUR SHUTTLEWOOD

ARTHUR BARKER LIMITED LONDON

A subsidiary of Weidenfeld (Publishers) Limited

Published in Great Britain by
Arthur Barker Limited
91 Clapham High Street
London SW4 7TA

ISBN 0 213 16713 1

Printed in Great Britain by
Willmer Brothers Limited, Rock Ferry, Merseyside

CONTENTS

DEDICATION

*As honeycomb cells are filled with the
essences of flowers, so words and works of
the wise are impregnated by the fragrant
essences of thoughts and aspirations. The
bees have departed, but the honey they
gathered and stored is left behind and is
sweet to our taste.*

*Gratefully, the author dedicates this
serious work on an important subject to
truth-seekers everywhere; especially those
who have heard, perhaps dimly yet insis-
tently, the Universal message breathed by
our Creator: 'The peace and stillness of
your minds forms my celestial garden.
Therein my flora and fauna of love blooms,
my fruits ripen, so to enrich the soul-spirits
of men ever entwined with My Being.'*

*I thank all friends and earnest research
students for their invaluable support and
encouragement in our mutual quest.*

Arthur Shuttlewood

December 1978

1

SLITHERING SERPENTS

On fairly rare occasions on night-shrouded hilltops in Warminster area, we have noticed bouncing balls of ruby-red light
that dance, bob about and bemuse the sense as they hare up
hillsides and emit a rising crescendo of sound. It is not exactly
a droning noise, monotonously pitched; rather, it resembles
a shrill whistle that varies in loudness according to distance.
They are sonic shapes, sometimes as circular as a cricket ball,
at others, elliptic and more like crimson spheres – yet often
pulsating in an irregular rhythmic manner.

Seen from far away, at a thousand or more yards, these
mesmeric spheroids are simply flickering fireflies of radiance,
reminiscent of sparks shooting from a wood fire, making illuminated patches in the night-darkened sky; but, when seen
through powerful binoculars, they are definitely brilliant, eye-
dazzling, miniature spheres which appear to be solid until
they swish through tree trunks and wooden fencing as if no
physical barriers can impede their silent progress.

Intuitively, we sense that they are associated with larger
aeroforms that perplex us and are quantitively lumped together as flying saucers or UFOs. Because we have studied
numerous examples of these minor specimens of mystic mar-
vels, we are sure that they cannot be collectively mistermed
methane gas, fireballs, atmospheric distortions, etc. Back in
1967, after a harrowing contact at close range with one of
these shrill-voiced objects, noiseless until it approached us in

the centre of Cradle Hill copse, we concluded that they were monitoring robots sent out by the 'parent' UFO. One cannot help suspecting, however, without tangible proof other than visual experience, that they are in fact flying saucers themselves, capable of changing shape, size, dimension and even character at will.

They may give us reason to doubt that all UFOs originate from populated worlds outside the orbital path of our own planet in a physically vast and immeasurable universe: they seem so much 'at home' as they spin and whirr dizzily close to the soil, usually weaving in and out of trees and bushes that cover the landscape. The sounds, staccato sonic high-pitched burpings, are distinctive; the air is alive with the almost shrieking cacophony when they swarm nearer to parties of earnest sky-watchers assembled on hill slopes, bewitched by bevies of such bouncing bubbles of flashing brightness heading for and entering into the hills before disappearing from vision.

Shimmering spheroids and dashing discoids that make a noise, yet sometimes cavort cunningly without a whisper of sound during their momentum around us, are not restricted to the Warminster region, my world-wide research tells me. At the time they first erupted over and around local hills, we thought they were unique to the tiny Wiltshire town that has become a veritable Mecca for UFO enthusiasts from all sectors of our globe; so it is pleasing and quietly reassuring to get confirmation of peculiar incidents from other sources.

For example a ball of white light that uttered 'a slight whistling sound' as it shot across the sky over Sidmouth in Devon puzzled many early-risers one Thursday a few years ago, the event well publicized in local newspapers. They tell us that some trawler-men were among those claiming to see the strange skyform at 7.20 am just as the heavens were lightening. Another witness was John Pratt of Church Street, who attested he was near Market Place when he spotted the ball of light whizzing overhead to vanish behind Knowle. He described it as similar to a magnesium flare initially, but added: 'When a plane drops a magnesium flare, it either falls straight down to earth or floats down gently on a parachute. This thing was moving at a terrific speed across the sky. I watched it for about 20 seconds before it disappeared. As it travelled through the air, it left behind a trail of vapour or white smoke about 1,000 feet up. At first I hesitated to talk about it because people might think I had imagined it, but I have since met several others who saw it.'

It *does* make a world of difference when fellow testifiers can support even the most bizarre UFO sighting story, perhaps elaborated upon, yet rarely embellished unnecessarily. We at Warminster, probably the biggest UFO 'hot-spot' on our native planet, are terribly conscious of this vital factor in relaying absolute truth. Sensibly trying to study and evaluate UFO evidence, one has to be an expert detective and sound sleuth to achieve any results, or at least proficient in the art of intuitively sorting out the gold of truth from the dross of lies in what we assess. When the two men met in 1920, Lenin said to H. G. Wells, eminent historian, novelist and writer of science fiction which often proved true: 'All human conceptions are on the scale of our planet. They are based on the pretension that the technical potential, although it will develop, will never exceed the terrestrial limit. If we succeed in establishing inter-planetary communications, all our philosophical, moral and social views will have to be revised. In this case, the potential – become limitless – would impose the end of the rule of violence as a means and method of progress.'

In agreement with this profound observation, I would concede that anything and everything which we, in our current scientific and technological concepts, think impossible, could be wholly feasible somewhere else in the gamut of universal structure where life exists. If electrical impulses can remain static in Salisbury Plain areas and still preserve original messages relayed ninety miles away five months previously (as related in my book *The Flying Saucerers*), to be re-employed in their entirety after such a period and lead

3

to military confusion, who can say whether other 'messages' of greater antiquity persist in these or similar areas throughout our Earth globe, somnolent and untapped, until disturbed? If they *do*, the planet itself a live recording unit, it is conceivable that words broadcast hundreds or even thousands of years ago are potentially capable of being harnessed today given that precise calibratory instruments and special frequency ranges are available to 'snare' ancient truths to propagate for future enlightenment. Stranger things have been reported on the perplexing UFO front!

In our attempts at Warminster to assemble and solve pieces from a great jigsaw puzzle of alien brand, we can but record what has transpired in the region and caused many furrowed brows and gasps of amazement. Take a few isolated examples that tantalize our efforts to compile that immense puzzle correctly: when a UFO cavorts and careers around in the sky at crazy speed, with stilted rhythms and lightning-fast changes of directions, observers at Starr, Cradle, Cley and Lords Hills are aware of unseen 'presences' at ground level. They are amorphous intrusions into our world of normality.

With increased regularity over the past few years, we have heard and seen such unworldly accompaniments to actual UFO sightings as the heavy, plodding footsteps of the Invisible Walker, only a few yards from us and disconcerting to nervous folk, non-traceable and non-catchable; the two-noted trilling of what has been termed the Tin-throated or Mechanical Bird, stereophonic and all about surprised sky-watchers, no naturalist or bird-lover able to identify the ear piercing warble of the unseen cosmic songster; and at quiet times of nocturnal peace, after a blazing UFO has faded into oblivion, the advent of three 7 ft tall figures, transparent yet clearly outlined, standing in a triangular or pyramidic formation at the bottom of the lane at Starr. They do not speak or communicate; and to date have been seen by some five dozen people, not all of whom would lay claim to being unduly psychic or to any degree especially attuned spiritually. One research student, bank cashier Neil Pike, walked

4

over to one of these silvery giants one night and began to talk to the tall figure with its gangling arms and domed head. In fact, eight of us saw him walk right through the glistening giant! It was as if Goliath and David in modern guise were captured by the camera of our eyes. Their movements are even, gliding motions, and they have no discernible legs. An exotic fragrance exudes from and permeates the air around them; and even on cold wintry nights the atmosphere is warmed by unusual heat waves from the areas where the trio congregate, stiffly erect and silent as thermal-type heat is generated from them.

Hard to believe? Savouring of ghosts and untouchables? Of course; but there are too many reputable and reliable witnesses to deny their reality, whatever or whoever they may be. Maybe they are, in essence, shades and shadows, entities from *future* time. Time is relative to the rotational and orbital speed of our own planet; it may not have universal application according to man-made calendars, the splitting of hours into sixty minutes, of minutes into sixty seconds, progression of weeks, months, etc. So surely to postulate that ufos and the intelligence connected with these aerial apparitions may indeed belong to a different space-time continuum from our own (may hail from the future, rather than the past) is not so ridiculous!

One clue is furnished; when the headlights of a car, or even the wan glow from torches, strike the apparitions, they evaporate like so much mist. I recall Terry Hayes (one of the camping party at Stonehenge in 1977) telling me that, as the ufos swept into view and were filmed by his male companion, Army searchlights tried unsuccessfully to 'fix' them in their probing beams. The searchlight rays were warped until they simply bent around the ufo craft and did not actually light their fuselages with a steady beam that would reveal the composition of the night-flyers. Terry mentioned an exotic floral perfume that lingered in the air after the Stonehenge ufos left the scene, too. At Starr Hill, Warminster (Middle Hill on Ordnance Survey maps), this is a noticeable feature; and the lovely scent of flowers that do not grow there vanishes whenever the shining figures leave the site, or whenever a low-flying UFO says farewell to the area.

Another mystery, perhaps the major one of all perplexities, stems from 'slithering snakes or serpents of sound'. They have an odd electrical crackling role, generated from underground. This subterranean 'creature' coils and uncoils, winds and unwinds, as it unerringly heads for the feet of observers standing in fields behind the barn at Starr Hill. They are nerve-shaking yet harmless, these peculiar serpents or dragons that swish and whine beneath the feet. It is an alarming experience for the victims; yet, more perhaps than any other unworldly criteria of the spectcular unknown, they indicate life that, although alien to us, might have been familar to ancient ancestors in the long, long ago. Maybe some fine day or night, they will cohere and make sense so that all will be revealed at one of these mass-vigil centres in our countryside?

Let us, however, make a more detailed survey of this sort of occurrence. I feel I shall be cheating the intelligent reader, or accused of withholding information that may be of value in future to other serious investigators, if I fail to record it and sundry revelations on a local front, no matter how far-fetched and extreme they sound. Much of the really spine-tingling aspects of UFO research of a deeper type began to manifest themselves much closer to the ground at Starr Hill in early February 1975. Were these weird phenomena linked with UFO intelligences? Or were they connected with a force that lies underground, within the very outer crust of our native planet?

I leave readers to form their own ultimate opinions and conclusions, but here is an absolutely true and typical example of what currently confronts us on heart-catching occasions at Warminster. There were sixteen sky-watchers present on a fairly clear February night on the roadway in front of the barn. Apart from two noisy and obvious planes, their engines punctuating the air with a deafening roar from

the echo-chamber formed by surrounding hills, one possible
UFO was spotted. Its aerial path was smooth, not jerky as
normally, yet its lighting was not consistent. It was most
erratic. It certainly did not appear to be a conventional aircraft, having blue lights twinkling spasmodically and not
regularly. However, it was pretty distant from the observers
as it skirted silently above the line of the hills, from west to
east.

It vanished from sight twice before its final blink-out
across the farm land fronting the barn. Shortly after this
came a harrowing experience that three of the watchers will
never forget. I was talking to two Warminster youths, in-
dustrial welder Alan Poolman and Andrew Riddle, both of
Woodcock Road, when we heard a light yet rhythmic thumping sound coming from the far end of the barn. Earlier, Alan
had had a near-encounter with the Invisible Walker, in a
field behind the building. For a man who up until then had
never seen a UFO – he was a newcomer to watching parties
– Alan was not scared of this rather harsh 'baptism' to un-
natural phenomena at soil level. In fact, he related to fellow-
watchers how he chased the unseen 'culprit' up a track until
the thumping sounds ceased at the edge of Battlesbury.

Now, a half-hour later and not long after the passage of
the aerial object with inconsistent lighting, he wondered
whether the thumping noises came from a stray beast or
marked the return of the heavy-paced and by now familiar
Invisible Walker, with whom he was anxious to get to grips.
Alan Poolman is a big, strong lad whose nerves are sound.
Several others at the vantage point had heard the heavy
footsteps too that evening, both before and after the possible
UFO sighting. There was also the infrequent and jarring call
of the warbling songbird that no one can identify, nor place
precisely. But Alan, Andrew and I walked across thick
patches of grass and weeds, past patches of scrub and a huge
heap of fertilizer *en route* to the rim of the ploughed field
behind the farm building. We each carried a flashlight. The
torches cast a trio of tongues of light over the ridges of soil

in front of us as we stood facing the gaunt shoulders of Battlesbury, early Iron Age fort and encampment.

We had suffered deluges of rain the previous month, so the soil was wet and soggy in the field. Then came a most peculiar happening: an experience we shall ever share but never forget! Perhaps – and this had often been noted, this recurrence of three – it was because a trio was present. Bob Strong, ex-RAF bomber crewman, housewife Sybil Champion and I formed an investigating trio in early research days at Warminster in 1965, the three of us undergoing some quite uncanny and dramatic ordeals, outlined in other books of mine on this intriguing subject. Few would believe us afterwards, so we learned to keep silent about the more bizarre and possibly frightening events. And now three of us faced a further ordeal for which we were totally unprepared, but at least witnesses were nearby.

As though attracted to the combined pencils of light stabbing from our three torches, we heard an eerie crackling noise emanating from the rain-soaked soil before us, about ten yards or so ahead. It came towards us from the heights of Battlesbury. It was somewhat like static on a radio set – with a distinct electrical element involved in the creeping carpet of sound. It was much as if a serpent was uncoiling and thrusting its sinuous length across the ground. A sibilant snake! We saw no rise and fall of the furrowed soil along its inclines, but the humming and hissing sounds came directly for us at about slow walking pace. The eerie crackling came nearer and nearer and we were thoroughly unnerved when it came right up to where we were standing at the edge of the field, muffled against the cold and wearing gumboots.

We instinctively jumped aside when it was a few feet from our toecaps, then we hurriedly took up a fresh position to the right, near the mouldering pile of manure about thirty yards away. The sound stopped momentarily, then changed direction without lessening its impetus and headed again for us, unerringly. The electrical crackling and sputtering was quite frightening, for here was something we had never before heard or been conscious of, in years of local sky-watching. Accompanied by a weird whip-cracking, or the sound of a flailing lash descending at speed, the slithering snake of sound, after curling its invisible loops of skin and as though baring its fangs, pronged towards us once more. Sensing it at our feet, seconds later, we ran this time to a clump of bushes near a running water-tap and length of hosepipe. The crisp crackling ceased and we thought with relief: 'That's it, then.' Already these macabre events over no more than fifty yards of rain-splashed ground constituted more than an average nightmare. But the respite did not last long.

This time it was like the demonic crack of a giant thunderclap. The noise beat at the sodden soil, seethed over the grass and whined towards our feet, hissing round our toes. Our nerves were shattered in that awful split-second of contact and we ran, unashamedly, to the hard standing of the road surface where our sky-watch companions were assembled. I have since learned that other small groups at this observation point have suffered almost the same kind of ordeal. Some Suffolk printing apprentices, who spent a night at Starr, heard the Invisible Walker pacing around their vehicle at four o'clock in the morning. They even tape-recorded the ominous footsteps belonging to no denizen of earth. Mrs Sally Pike, daughter of a retired detective superintendent of police, living at 14 St John's Road, Warminster, even saw the impressions of the heavy walker – no body attached – in a large pool of water lying in a depression outside the barn! Incredible? Impossible? On the surface, yes – yet things verging on the miraculous have indeed transpired at that notorious UFO-haunt, Starr Hill, Warminster.

Are the warbling mechanical bird, the heavy-stepping Invisible Walker, and the crackling serpent all unearthly derivatives from the same source? And do they all, even remotely or indirectly, tie-up with our worldwide UFO mystery? But back to that frantic February night of 1975. Recovering our composure after the steadying influence of piping-hot coffee and over-loud assurances from more phlegmatic among fellow observers that 'nothing can harm you', Alan and Andrew later ventured into the equally rain-soaked field on the opposite side of the narrow lane. I fixed my torch on a metal gatepost, where it automatically blipped amber flashes of light. It was only when their own torches flicked out rays of radiance in the field that the crackling phenomenon renewed its 'attacks'.

This was seen by all the co-observers, who can testify to the strict authenticity of the experience. The two lads were thankful, in that morass of mud and slime, that they were wearing gumboots. This time, the 'viper' showed that it had no sting or animosity. There was no harmful intent in its mind-gouging antics. It went right up to where they were standing, or slipping, in the mud-coated field. They heard it crackle and whine up to their very feet, but this time they stood their ground and bent down to touch the damp soil in front of them. The electrical sounds stopped immediately, while the rest of us stood gaping and fearful that the lads would be electrocuted. Some days later, fearing that they might have been exposed to the dangers of an electrical current of unassessable strength, I asked them how they felt. 'Fine, just fine,' enthused Andrew. 'We are looking forward to the next meeting with the slithering serpent!' Alan nodded his agreement. 'If UFO sightings are as exciting as what I went through last Saturday, you won't be able to keep me from Starr Hill in future,' he grinned.

Not many Britons know that Thor Heyerdahl, of *Kon-Tiki* fame, who crossed the Atlantic from east to west in a replica journey of an ancient Egyptian papyrus boat called *Ra II*, was among hundreds of people who witnessed a UFO reported from a number of Caribbean localities on Monday, 29 July 1970. The story was given no prominence in the English press, yet was the subject of a lengthy item in the Norwegian newspapers of 1 August. Obviously, it was a news story of some importance, so merited the headlines accorded it. In brief, the story ran:

10

San Juan (NTB-UPI) – Thor Heyerdahl and his crew on board *Ra* II, the crew aboard the United Nations oceanographical ship *Calmara*, and thousands of citizens at St Thomas, St Croix and other Caribbean Islands, reported on Monday night at 0245 that they had observed an unidentified flying object.

The first report came from a navigator on board *Ra* II, Norman Baker, who said that he had charge of the steering of the vessel when he caught sight of the 'thing' which was round, flat and clearly lighted. He cried out to Heyerdahl and the Mexican anthropologist, Dr Santiago Genoves, who was with them. All three looked at the 'thing' for ten minutes. Also *Calmara*, which was on course towards *Ra* II, reported that they had seen the object. Newspaper and radio stations on several Caribbean Islands reported that they received hundreds of telephone calls from people who had observed the 'thing', also many fishermen out on the sea.

That was the story, which created quite a sensation among Norwegian readers and radio fans; but, so far as I can gather, a few scanty lines was the total output of news from the British media. Thor Heyerdahl is one of the greatest adventurers and explorers in the world.

Even the most hard-headed realist will agree that, when shrewd observers of the undoubted calibre of Heyerdahl and Genoves, both trained scientists, plus a professionally trained crew of an oceanographical research ship, report seeing a strange and baffling object in the air, there can be little doubt that not only was such an aerial apparition seen but it was sufficiently odd for special note to be taken of it. This was a first-class report, judging from the hosts of witnesses, and one that could well go down in the annals of Ufology as a classic sighting. Yet must we ignore or disparage the testimony of the not-so-famous? Their evidence, frankly given, is as soundly based as that coming from the most intrepid sea voyager.

2

THE NATURAL PATTERN
DISRUPTED

When the message reached me on the eventful evening of 13 April, 1968, it started a staccato rhythm of excitement jabbing through my whole being, my every fibre and sinew. Yet I recall the short laugh with which I greeted the caller, who was insisting I should jump into his waiting car at my door. It is as well to have a sense of humour, even in serious research work, for nothing is so disconcerting as to disbelieve a seemingly incredible happening, and then to have cosmic truth forced firmly down one's throat by a stunning repeat performance that indelibly stamps the mind with the brand of irrefutable authenticity.

Bob Strong, then my chief aide in investigation of the spectacular Unknown in local skies and an ex-RAF bomber crewman, called on me around 9 pm on that memorable night in Warminster. His brown eyes bulged under thick brows as, gasping with ill-concealed excitement, he blurted his extraordinary tale of a near-contact with a giant, swaying UFO. I turned away to hide a sceptical smile. It was plainly too outrageous to be true! Or so I thought at first as he ushered me into his car for a return trip to Cradle Hill, as he put it: 'To see for yourself.' I found it difficult to credit the sensational story he jerked out *en route*. It was just too crazy to be true! About an hour earlier, he told me soberly, he had carried out a short solo sky-watch at Cradle, by the white metal gates. He was surprised to see a fiery-red object plummet downward from the heavens directly overhead, rapidly changing from a crimson cone of light to a vast rectangular aeroform the size of a 10p piece at arm's length.

'It veered unevenly and yawed over to the skyspace above the golf clubhouse, hanging vertically in the air but swinging from side to side, like a drunken balloon with blunted ends,' said Bob quietly. There was a palpitating, fluttering movement from the oblong shape as it swayed like a clock pendulum in a fitful falling leaf motion for several minutes. Then it reverted to a horizontal aerial stance and sped to a spot over the nearby copse, ablaze with a roseate glow. That is when, according to my companion, three RAF jet fighter planes zoomed across the sky with a high, screaming whine. Fighter aircraft carry blinking belly-lights which flash redly and fairly rapidly, pulsating regularly as they hurtle in for the 'kill'.

These three maintained their bright pulsations as they moved in with obvious intent. At the advent of the fighters, however, the Thing shot up into the upper atmosphere like a tracer round speeding from a machine-gun! It eluded its pursuers by hiding behind a thick belt of dense grey-white cloud, while they split up from formation and flew in differing directions, hunting their prey. The UFO then showed that, whatever intelligence was aboard and in control, it possessed a delightful sense of fun and enjoyed its enforced game of 'tag'.

In absolute harmony with the pulsing belly-lights of the three jets, with split-second timing that synchronized perfectly, the UFO belched forth a vivid corona of light that gaily and garishly girded clouds affording it cover. It was as if it was deliberately designed to assure any watchers from ground level that not only was it acutely aware of being pranged by the trio of jets, but it also wished to prove to all and sundry that it had the human capacity for playful imitation or humorous impersonation! This portion of Bob's amazing story I stalwartly refused to credit as factual, yet

13

stranger things were to happen at this later hour as we proceeded to our destination.

When we arrived at Cradle Hill and stood at the top of the road, Bob finishing off his astounding narrative and I still turning aside to smother my half-smile, something flashed redly over the clubhouse. It riveted two pairs of eyes and we were treated to an almost precisely patterned 'repeat' of the previous aerobatic performance in the heavens. It was now glowing pinkly with the onset of sunset and the effect of lowering cloud layers. Before the jets came onto the aerial scene this time, we noted that the swaying structure of the stranger was gradually wending its way over the copse, bouncing along the tips of the trees as if intending to drop and effect a landing on the field beyond.

Doubly-impressed Bob and the author rushed up the rough track towards the clump of trees, thrilled to think that they had made a definite contact with the elusive UFO intelligence at long last and that they would soon meet the occupant(s) at ground level instead of merely witnessing colourful aerobatic behaviour during all-night vigils. Such a welcome was not destined to take place. As the bloodshot giant resumed an even keel, bouncing and bobbing up and down beyond the treeline, a trio of jets screamed harshly overhead, coming either from RAF Boscombe Down or Old Sarum, we guessed. We can but surmise, for very few in authority among our Armed Forces will dare admit openly that unidentifiable craft have been vainly chased by aircraft of more conventional types on our planet. Now I simply *had* to believe Bob Strong's original story spun in a peaceful April evening, for the crimson aeroform immediately took shrewd evasive action!

It skimmed aloft at blistering speed and I was conscious of the fact that two trees were bent over slightly at the copse, as our visitor blasted off soundlessly into space. There was not a consolatory whisper of noise from the departing UFO, even when the jets separated and milled around the golf course area in their fruitless search for the intruder. Their

14

quarry once more sought a wispy blanket of cloud as protection against the trio, and its winking red glare spurted
from high over our heads, flashing in perfect unison with the
belly-lights of the frustrated jets. Bob's unbelievable story
was amply vindicated! The planes flew off after four minutes,
disappointed no doubt, and we waited expectantly for the
return of our high-flying space-user. We were there at the
white gates for quite some time without reward, for it did
not reappear from the patch of dense cloud, which slowly
thinned and became golden lances in the harmless armoury
of a blazing sunset before ink-black darkness dominated
everything around 10 pm. Still, we waited. At last, a star-
speckled night sky swallowed sun and clouds from view, and
we were left with a precious memory only. A vivid remem-
brance, moreover, which still tickles my fancy and sense of
fun, so many years after it happened during the course of
the Warminster UFO saga since the Christmas of 1964. Here
was patent proof that not all unknown celestial chariot crews
are devoid of an obvious sense of humour. In this particular
case it was definitely outsized! Yet is it conceivable that
there is a dormant clue in what took place that eventful
night? Could the intelligence have been mutely and pic-
turesquely stressing: 'It is nothing new. There is nothing new
under the sun. It has all happened before in Earth history,
many times . . .'

Whether they are extra-terrestrial or other-dimensional,
hailing from outer space or nearer to our planetary environs,
what is the UFO intelligence trying to warn us against? Ghosts,
phantoms and spectres may be shades of the past. Time,
however, is a man-made concept in a scale pertaining to the
Earth planet and is *not* universal in its application. So – a
novel thought – could UFOs be loosely termed messengers
coming from our *future* time? 'We have come back in time
to help you.' Could this imply that certain very evolved
people are coming back from their own advanced time continuum into our own present one? So what may they be

mutely warning us against? Apart from the acknowledged dangers of nuclear proliferation and spread of radiation hazards and other social evils, there may be clues in a rather frightening series of incidents that tell us our birds and bees are in revolt in a wild, wild world.

Warnings came in the early months of 1975, when we learned that there had been swarms of killer bees in South America, attacks by birds on people in America and Poland, and a suicide onslaught on a small town in Illinois by rats. Flying ants ravaged much of Australia and giant snails appeared in Florida. Marauding packs of wolves and lynxes struck at remote farms in Canada and Alaska. A 100-mile stretch of beach in California was invaded by swarms of gigantic squids. A school of bluefish savaged bathers off Miami Beach. Terrified swimmers staggered from the water, some with gaping wounds caused by the razor-sharp teeth of the Caribbean bluefish. The Miami Beach lifeguard chief said he had 'never seen anything like it before'. Two girls in sleeping-bags were attacked by a pack of coyotes in a state park in Missouri. Scientists in the USA viewed with concern disquieting reports from all over the world of strange behaviour by animals, birds, insects and fish. The inhabitants of Prince Edward Island in Nova Scotia spent a week under attack from five million crows. Even shotguns could not scare them away! What did it all mean? One theory was that man-made radiation, from such things as power stations, radio, television transmitters and even microwave ovens, is creating electro-magnetic radiation affecting animal life. One scientist, studying the effects of long-term exposure to electro-magnetic radiation on rats, said there was evidence that it can break down body chemicals and even affect the brain.

Another discovery is that aerosols can eventually affect the protective girdle around the planet, so their use has been discouraged. A report from the Soviet Academy of Sciences has endorsed USA findings. It suggests that changes may be taking place in the biological system of our world. A spokes-

16

man said: 'We are witnessing rare and unusual animal behaviour, strange migrations, and sudden population explo-
sions in insects and animals.'

When considering the origins and purposes of UFOs in our
atmosphere, almost anything is feasible. So, could these be
some of the unexpected perils on Earth that flying friends
have been warning us against for the past three to four
decades? The clues are all there: electromagnetism, microwave radiation, delusions and blackouts, prowling pumas and
sundry weird animal stories. Birds? What of our tin-throated
or mechanical warbler at Warminster? Just a thought, of
course; but could it be that someone is drawing our mass
attention, urgently, to something that has gone awry and
conflicts with perfect natural patterns in life?

An incredible true story of a ten year time lapse in de-
livery of a message was related on Harlech TV's news transmission from Bristol on 11 July 1978. Sailing at sea, the liner
Queen Mary II received it and the telegraphist on board
verified that it was obviously addressed to the *Queen Mary* I,
out of service for several years and whose call sign and code
signals were adopted by her successor, the second *Queen*.
It was a message sent from Portishead, Bristol, the telegraphist taking it down with customary skill and swiftness on
his message pad. However he was puzzled; and captain and
crew were soon acquainted with the astounding revelation
that the signal was intended for the first *Queen* – and had
actually been sent ten years before its eventual receipt by
the second *Queen*!

A message transmitted, yet 'lost in space' for a decade?
An inquiry was set in motion, but the answer still eludes
radio operators, scientists and technology generally. One
scientist, when interviewed on the programme, after a reporter had spoken to the dazed telegraphist aboard the
second *Queen Mary* and confirmed the electrifying statement, gave his opinion that by some quirk the message had
been relayed out of Earth space, gone from our atmosphere

17

and was snared by another planet; then ultimately, after travelling for ten years in the voids of deep space, had returned to our Earth planet and been picked up by *Queen Mary* II's receiving apparatus. Crazy? Of course it is. Worthy of anything listed by Charles Fort, collector of oddities unlimited.

One must concede it a possibility that, during those ten lost years encompassing transmission and receipt of the message, someone rather than something re-directed it on to its correct destination, or to the current code and call sign of the defunct *Queen Mary* I; which – again possibly – means that UFO intelligence was at work!

Another bizarre news story broke on 2 June 1978, a Wiltshire press organ telling us:

Possible sites of the Golden Ram of Satan were pointed out by Mr John Forward, of Longhedge Farm at Corsley, less than three miles from Warminster, when he led a public experiment in dowsing on Cley Hill on Wednesday evening of last week.

Using copper rods which answered his questions by twisting and turning in his hands, Mr Forward claimed that there was gold at a depth of 271 ft below Cley Hill. He said there are legends of a golden ram of Satan buried somewhere in Europe in a triangle of tunnels, and asserts that Cley Hill is in the centre of such a network, as pre-Reformation passages linked the great houses and churches of the area to enable monks to carry valuables safely. When Mr Forward asked the rods to show him the direction of the Holy Grail, they pointed towards Silbury Hill (near the huge stones of Avebury Ring). The local farmer believes it is buried there, and says that the BBC dig which revealed nothing several years ago must have been in the wrong place.

About fifty people followed Mr Forward around the hill on an overcast and windy evening; and for some time a

18

rather fractious herd of bullocks followed too. The public were invited to use copper rods for themselves, and many found the rods turning strongly and inexplicably in their hands. Some early readings, at the foot of the hill, Mr Forward attributed to underground springs. But stronger readings, he felt, were due to an underground tunnel. He told how he had taken borings at the spot and brought up brickwork from a depth of thirty feet. He thought that this was a spur tunnel from a much longer tunnel running from Longleat House to the Church of St Denys, Warminster.

He explained that he and other people had investigated parts of the area's tunnels, although large sections had been bricked-up or had fallen in. After he and his son had been buried up to their chests in a roof fall, his wife would not let them go down again!

That was the news story read by thousands, the farmer-dowser narrating it on Radio 4 of the BBC when interviewed. UFOs have often been seen in mystical motion over and around Cley Hill in the years since 1964; and several ley-lines cross this lofty eminence that is but two hundred feet short of mountain status. Could it be a twin to Glastonbury Tor?

Naval signals were picked up by Army receiving sets during an exercise at Imber, ghost village on Salisbury Plain; and the original messages were relayed from battleships on patrol in the English Channel five months prior to the Army battle rehearsal! The true episode was mentioned in one of my earlier works, and I have spoken to soldiers who took part in that particular exercise, confirmed by the School of Infantry commandant at the time, Brigadier A. Arengo-Jones. But ten years! In journalese parlance, it is a king of a story.

When one becomes entangled in unknown aerial tentacles of strange monsters of nebulous nature, consistently over a period of fourteen years, the investigating mind re-

19

mains baffled and bewildered by weird aerobatical antics of celestial chariots, luminous aeroforms that have haunted one particular small town so regularly in that time span. One either accepts the challenge of the collective enigma whole-heartedly in a spirit of healthy curiosity about universal structures and life-forms, or abjectedly 'throws in the towel' and acknowledges defeat in face of a great and spectacular unknown.

Thankfully, when the latter course, the easy way out, tempts the humble researcher to avoid a mind-bending and impenetrable maze and jungle, support is always at hand from earnest souls throughout the native planet. Many folk throughout the world have had unnerving yet exciting skirmishes with awesome creatures living in the waters of deep space that – abruptly and unasked – land on Earth from machines of unearthly design and aero-dynamical mastery. For, at present, an inescapable fact is that not a single human being on our tiny rotating sphere can claim to be a fountain of all cosmic truths in a vast, immeasurable universe; no mind can ever be a repository for complete knowledge and wisdom. That submission is meekly expressed and undeniable.

Fortuitously, some of us sincere seekers have been presented with minor fragments of the gigantic jigsaw puzzle that so-called Flying Saucers leave in their dazzling wake after incursions among us, unsolicited and unexpected. So we at Warminster, focal point of so many of these congregations of inexplicable and unidentified chariots of the heavens, feel it our bounden duty to fellow-seekers of eternal verities everywhere, to share freely our experiences of the most mysterious and probably most important conundrum facing mankind since the birth of Christ.

Final conclusions can safely be left to earnest students intent upon delving deeply into the imponderables of these glowing ghosts in our skies: the Who? What? Where? and Why? questions that gnaw at our minds incessantly. Warminster hilltops, some two miles north of the town, which are most frequently visited by vari-hued spacecraft or shimmering skyforms, are on a direct aerial route between two centres of ancient magnetic power – Stonehenge to the east, Glastonbury to the west. Whoever or whatever the controlling influence and guiding intelligence behind these alien incursions into our hallowed and sacrosanct air space, careful vigils maintained at these crucial hill areas prove that UFO flights occur most often along this line. Viewers of Granada and other ITV networks saw some remarkable colour film of UFOS hovering over Stonehenge in the spring of 1978. They were taken by one of the two men present. One cannot fake movie film, and there were about four minutes of it on the reel. The photographer, who was camping on the outskirts of the prehistoric stone circle for three days, saw no fewer than eighteen positively identified UFOS during the three nights of sky-watching. He did not wish to have his name revealed, but Terry and Valerie Hayes, of Cheshire, told me the senses-tingling story at first hand, shortly after their experience.

Terry, a carpenter-joiner for British Rail, and Valerie, nursing midwife, were truly staggered at their sightings; gleaming UFOS on one night remained flitting and floating over the stones for some three hours. Their near-encounter will be fully described later in this work; and I trust to have some 'still' photos taken from the priceless film that Terry Hayes's friend took in the small hours during their event-filled sojourn at Stonehenge; for the craft depicted are precisely what patient watchers at Warminster hilltops have seen over the past fourteen years and more.

Stonehenge is only some ten miles from Warminster, as the crow flies. Many ley-lines (also described in the next chapter) cross this ancient and time-worn circular monument where the druids still continue their annual celebrations of ageless rites, strongly associated with natural forces that pertain throughout our wonder-filled universe.

21

3

MAGNETIC PROPERTIES
OF LEY LINES

Many curious people who visit Warminster on sky-watch expeditions ask our research team what we think about leys. And what, precisely, are leys? My colleague Paul Screeton, editor of *The Ley Hunter*, is better able to deal with this perhaps pertinent query than I, so let us learn what he has to say on possible cosmic energy links betwixt earth and sky. Once the sparetime occupation of a ridiculed minority of archaeologists, the fascinating pursuit of ley-hunting had almost vanished until it received a boost from an unexpected quarter. Alfred Watkins, who pioneered research into alignments of prehistoric antiquities and old parish churches, believed the leys, as he christened them, were purely to mark trackways. Maybe he had an inkling that they were something more, but he and his fellow enthusiasts in the Straight Track Club concentrated their efforts on collecting evidence to support the theory that prehistoric man travelled everywhere in straight lines.

Yet it was a club member, Major F. C. Tyler, who postulated that many sites were also located on important sites such as Stonehenge, and found that the radii of these circles were all multiples of a prehistoric unit of length discovered by fellow club member Arthur Lawton. It was Lawton, it seems, who first related leys to a grid of 'power' emanating from the earth, these radiations being the clue to the real purpose of the ley system. An attempt to link leys and orthotony, probably stemming from 'straight line' UFO flights, introduced a whole new line of research and a whole new sector of the public to the subject. Orthotony, as most Ufologists are aware, is the theory that UFOs fly in straight lines, and Aimé Michel propounded the theory that one day's sightings of different UFOs travelling in different directions can be plotted on a straight line. John Michell, a fellow-author living in Bath, points out: 'I think most of us realize that the phenomena of UFOs and prehistoric alignment are in some way related. It was the reports of UFO sightings that led to the present re-examination of the ley system.' Alfred Watkins was years ahead of his time, and had it not been for Jimmy Goddard and a few others, who some years ago suggested an association between leys and UFO paths, his work would have remained neglected.

Unfortunately the association is very hard to prove. The ley system stretches all across the country, set out with the accuracy of a modern surveyor. The precision of alignments between individual standing stones can only really be appreciated on maps of the six-inch scale. On the other hand, very few UFO reports contain any exact information of the location and direction of sighted objects. In general, UFOs and migrating birds do appear to follow certain lines of exceptional magnetic properties such as geological faults. The earth's magnetic field, and everything within it, is directly influenced by the sun, moon and planets; and this may well explain the prehistoric practice of astronomy in connection with leys and alignments. The legends of dragons as manifestations of cosmic energy indicate that the phenomenon of UFOs was known to prehistoric astronomers, as it still is to primitive peoples, and that its nature was once far better understood than it is now.

John Michell says that is why he believes that the most rewarding approach to the problem of UFOs may be through the study of prehistoric science and philosophy. But if we do accept that there is a connection between leys and UFOs, then we must ask 'How?' and 'Why?' Tony Wedd, a Kent artist and designer, believes the leys were used by UFOs to pick out landmarks, but Paul Screeton considers that the obvious sophistication of these craft and the exceptional speed at which they normally traverse our skies makes this extremely unlikely. However, the answer could lie in the 'power' emanating from leys. Possibly some, or all, such craft use this in some manner, either solely for propulsion or for direction-finding on major leys. They use leys blind as if homed on the line, rather than looking for and navigating along them.

This is wholly speculation, as the precise nature of the power is not known – neither is the method of propulsion used by UFOs. Yet Warminster is a remarkable ley centre and, of course, a prominent UFO locale. Another suggested link between leys and UFOs is that prehistoric man, when plotting the ley system and marking it, received help; that is, help from the occupants of UFOs. Historians have recently discovered conclusive evidence that mankind was in contact with UFOs in prehistoric ages. What is less easy to determine is the nature and extent of their aid. Did they educate prehistoric man in the sciences of geometry and astronomy, specifically to enable them to map the leys; or did prehistoric man, possessed of more intelligence than we give him credit for, do this unaided? Is it coincidence or imagination which makes Stonehenge look from the air like the conventional disc-shaped UFO, even though the work of Professor Hawkins, Boyle and Thom has revealed the circle of stones to be a fantastic computer? And is it significant that the huge zodiacs laid out on the land can really only be appreciated from several miles up in the air?

So much speculation may deter the scientifically-minded and attract the crank, but we must not be afraid to put forward such ideas, provided we make it sufficiently clear that it is purely speculation and not concrete fact. Alfred Watkins was mocked for his pains, but he staunchly continued his researches, even though those who denigrated him have been found to be partially correct. Many leys go over precipices or through bogs, and if they were simply tracks they would make far from easy routes for a traveller! It seems that the use of leys as paths followed their prime purpose as markers, not simply because it was convenient to go from one marked spot to another, or that their original purpose had been forgotten, but because the 'power' is beneficial.

Their use was not forgotten until much, much later than the Roman invasion, as evidence shows that churches, probably to pre-Reformed times, were built on them, showing that some knowledge of their special nature and sanctity was remembered. Nonetheless, there is little known about the nature of leys, and it could be a long time before they become academically acceptable. A mathematical probability theory has been worked out by Tony Northwood, of Bedfordshire, and work is still needed to apply it to leys throughout the country, so as to prove beyond reasonable doubt that leys exceed probability. In a case where he used the formula, he found the chances of six sites falling on a straight line to be one to many billions. At Warminster there is a twelve and a thirteen point ley! This approach proves alignments exist yet gives no indication of their nature. For this we need a form of divining – dowsing – which will show where leys are and indicate their character to a 'sensitive' person. Further, no fewer than twenty-three complete constellations of stars are noted in Southern Counties, from chief barrows, as seen from the air.

Certain properties of leys have been propounded, including the supposition that they encourage fertility of the land: that at certain times of the day a sound can be heard along leys; that they have recognizable hues of colour; that they have geographical forms slightly different to their immediate surroundings; and that at particular power centres a buzzing is felt in the head. To the archaeologists the leys are fascinating because they suggest that prehistoric man was

B

highly advanced scientifically. To the Ufologist they are equally intriguing because they may hold the clue to the mystery of the flying saucers. All I can add as a researcher is that this is an oddly compelling survey of leys, or pre-historic straight lines and tracks that exude possible power. I can verify as far as Warminster is concerned that UFOs do indeed seem to follow specific aerial routes more than others – between burial barrows, tumuli, springs and fountains, original water-courses, former sites of pagan worship, etc. – also, that close proximity of the 'machines' over Celtic ridges and barrows appears to have a strange effect on some sky-watchers. (The usual aerial path over Starr Hill lies between Glastonbury and Stonehenge, two enigmatic sites.)

It is as though a catalyst explodes, so that members of the group beside the barn at Starr or near the central copse at Cradle begin to see weird shapes and even hear unusual noises; they become silent, introspective, deep in thought, experiencing a mild giddiness. But only some of the observers are so affected, not all. Perhaps because they are more sensi-tive to the changed atmosphere when UFOs are near? Who knows? It is as if currents of special energy are discharged from ground level and matched by some vibrational counter-part in the sky, with the incredible UFOs 'neutralizing' and using the admixture of the two forces, in between. But again – who knows?

Library worker Wendy Channell of Graham Road, Blacon, Chester, tells how a regular came in one day to ask if any books on UFOs were available? 'This was a change from what he usually read, and it turned out that the reason he asked was he recently returned from holiday with his family in Minehead, Somerset. While there, they all saw peculiar sky objects late at night.' Here is the report made out by regular reader Mr J. N. Ironside, once a sergeant paratrooper. He is naturally a firm believer in disciplined facts and not fancies:

There were other people present and I remember them all pointing upwards. What we all saw and marvelled at were

26

nothing like aeroplanes. My daughter saw them first, thinking they were shooting stars. Looking up, I observed three large circular objects at an estimated two thousand feet.

There was no noise. Each had a set of lights around it. One then moved east. As it did so, it tilted on its side. The lights revolved faster, until it was one continuous circle of light. It moved very slowly for about two miles, then turned and moved back. Then it joined the other two, and the three skycraft soared up at fantastic speed. Within four to five seconds, they were lost from sight! I estimate they went from zero to well above the speed of sound within two seconds. But still no noise. My wife and two daughters were with me, it was between 11.30 and 11.45 pm, and weather conditions were: no cloud, still, and very stormy. We shall not forget our holiday in Minehead this year – not ever! Other folk were there, but I couldn't hazard a guess as to how many. All gaped skyward, anyway.

It was no earth-made machine that caused considerable discomfort to businessman Leonard Pyke when he was motoring to Frome from Warminster on an early September night. He told my co-observer Sybil Champion that the interior of his car became unbearably hot and his ears were shattered by the almost deafening buzzing of millions of bees. It filled the inside of the vehicle, bemused his senses, heat growing more intense all the while at that notorious spot on the road near Cley Hill, Corsley. The phenomenon terrified him! His motor began to cough and whimper, shuddering in the few yards it took before his brake clamped hard. He jumped out of the car, which shook as though travelling a bumpy byway, although its engine was dead.

Gazing up, he beheld a large white circular aeroform spinning overhead. It weaved in a jerky motion from side to side directly over the trembling car and its driver. The craft banked slightly as sound tore at Mr Pyke's eardrums. It went into an upward and spiralling movement, did a figure of eight, balanced on its rim before becoming temporarily elliptic, then rolled away. It hurt his eyes to keep looking at the twisting and turning spheroid. It eventually sped towards lofty Cley Hill, remaining stationary over the smaller of the two peaks for about forty seconds before darting away to the north and out of sight.

Several UFO-inspired incidents, late in January 1975, confirmed my view that these incredible aeroforms are quite capable of stopping car engines and even affecting power supplies. One concerned a couple who were driving past a Roman fort outside Bridport in Dorset when their engine stalled. The lights would not work, nor the ignition. Then they beheld a ball of blinding light hovering in the sky overhead; and when this spheroid moved off, the car started again and purred normally. The Bridport weekly newspaper reported the story, plus other untoward experiences in the area, for days afterwards. Another sighting, nearer Warminster, involved a man taking his dog for a walk across the downs at Hanging Langford, along the Warminster to Salisbury route. He became aware of a brilliant light that sparked the night sky directly above him.

The dog was absolutely terrified. It quietened when the gem-like UFO moved away into the distance like a duck bobbing against a fast-rippling stream. An alleged landing of a UFO in Dartington, Devon, was given a few lines in a morning newspaper. Electro-magnetic failures in motor vehicles are fairly common where UFOs haunt local hillsides. Bank employee Neil Pike of St Johns Road, Warminster, reported:

On the evening of Monday 21 October 1974, whilst sky-watching at Starr Hill in the company of Arthur Shuttlewood, an experience of these effects took place. My car is equipped with parking lights. These are positioned on each side of the bonnet and are battery operated. They had been inoperative for many months. As they are not a legal requirement in England, I had not attempted to trace the reason for their failure.

At precisely 9.5 pm the light on the passenger side of the car became illuminated. At that time Arthur and I were some distance from the vehicle. We returned to the car and I found that the switch controlling that circuit was depressed. With the aid of a flashlight we both examined the wiring. Then we wiggled the bulb-holder to ascertain whether there might be a loose fitting. All attempts to extinguish the light by any means other than the main switch failed dismally. After five minutes the light was turned off by means of the switch. We resorted to wire-fiddling and once more met with no success. Ever since, the parking light has not worked again.

So what, or who, caused this incident? It is another minor mystery to be written-off as inexplicable.

There is scant evidence, especially of a testable and tangible nature, that all UFOs originate from extra-terrestrial sources and that crews are of humanoid physique. Frankly, on the Warminster area assessment alone over the past fourteen years, I strongly doubt whether more than a small minority hail from worlds similar to the earth planet. Most belong to a normally invisible realm that interpenetrates our own denser sphere, peopled by entities who can materialize or vanish at will because they belong to a vastly different dimensional scale and existence level, even if of basically the same genus and species. That inveterate compiler of freakish oddities that seem wholly disconnected from natural phenomena, Charles Fort, said of the world-wide incursions of UFOs: 'I think we are property. I say we belong to something. That once upon a time this earth was a no-man's land: that other worlds explored and colonized here and fought among themselves for possession. But now it is owned by something; and that something owns this earth and all others have been warned off!'

He is not the sole shrewd chronicler and investigator of unknown incoming intelligences to conclude that the aliens are so at variance with anything we know that to try and describe them in language understandable to everyone is an impossible task. In essence, our owners, creators, and possibly our genetic controllers in the cosmos, are not merely from another world but from another type of universe, and their laws of chemistry and physics are foreign to ours. They can break time and space barriers, travel anywhere in the galaxy by hyper-dimensional links not yet within our own scientific horizons, and normally remain invisible to our circumscribed vision which fails to see beyond the spectra of the ultra-violet and infra-red boundaries.

Some psyche-dominated and spiritual thinkers believe they are pure thought energy forms and can be anywhere at any time they please. The weird corollary is that they can actually appear in physical guise when they wish; and in this atomic and molecular cell-changing manner have been responsible for nearly all legends of heavenly angels and great white gods from the skies which have been born in various countries and scorned in these so-called more enlightened times. Looking at certain UFO testimony and flight patterns in retrospect, some deep thinkers say Ufonauts are totally uninhibited by physical limitations such as fleshly bodies. Yet it is reassuring to us at Warminster, blessed or lumbered by so many UFO appearances since 1964, to have many corroborative sighting stories from other sectors of our native globe. They exonerate us from unnecessarily cruel jibes and cynicism.

Here is a quaint facet of the mystery . . . For about six weeks in early 1966, when the planet Jupiter was very prominent in the heavens above Cradle Hill at Warminster, we were puzzled by two unlisted 'stars' that were equidistant from the planet, Jupiter comprising the apex of a triangle. The flat yellow orbs we termed 'pinion' or 'anchor' stars, because whenever a UFO came over that particular span of sky at night it habitually encircled one or other of the two base stars. This aerial practice was so consistent, night after night, that I made an effort to confirm it when spending a weekend at the Chelmsford home of my parents in Essex. For six nocturnal hours I kept diligent watch from a high point at Danbury, a nearby village. I saw but one satellite in that period, but here is the astounding revelation in miniature that struck me by its contrary nature: Jupiter was still in the sky of course, yet there were no pinion or anchor stars below it forming a triangle, as there undoubtedly were over Warminster. When I arrived back, I immediately contacted my team-mates, Bob Strong and Sybil Champion. Yes, they assured me, the peculiar base 'stars' were still present during night hours. A number of astronomers and UFO enthusiasts belonging to BUFORA (British UFO Research Association) will corroborate that these stars were seen by them although unmarked on any sky map.

After six weeks of dimly glowing over our noted hill, the yellow orbs suddenly flared into unmatchable brilliance one early May night, and simply rolled away to disappear farther aloft in an elegant sweeping motion, just as if we had been imagining it all! Deduction? They were space stations organized by our visitors for a specific purpose, apart from causing a stirring of our grey matter. 'We have made your town a base for our operations,' said Aenstrian callers in my first book *The Warminster Mystery.* You will remember that we did not infer from this that the base was an aerial one rather than established entirely on terra firma.

Heat, weird humming, stopping of an engine: these features are consistent with scores of cases similar in nature around Warminster area, and, since, all over the world. Do these disturbing happenings denote hostility, or merely a means of shocking fresh 'victims' into realization that other-worldly forces are at work in our atmosphere, whether we choose to welcome them or not?

Propulsion media? Data concerning unusual magnetic

fields can be attributed to; ionization, colour, sound spiral vortex, latent energy in the atmosphere, as well as many other things all based on nature, so perhaps it is natural forces we should be studying more minutely for essential clues. Telepathy should also come under sympathetic survey. Experiments are being carried out already in this field; crews of long range atom-powered submarines are being taught rudiments of telepathic communication, which would have been scorned in pre-war days. Telepathy cannot be picked up by radio-scanners.

People already claim the ability to pick up thought waves or ether vibrations from others far away. There have been a number of quite stunning instances of disasters predicted by persons nowhere near the scene of the tragedy. Pre-cognitive experiences and fateful dreams are recorded in history, yet sadly remain a great perplexity to science and leading thinkers. Science has long abandoned use of the word 'ether' in describing properties of space, only to discover, gradually, that it was not perhaps the simple dream in the minds of William Crooks and his contemporaries in the last century. The theory of Plantier has been analysed pretty shrewdly by several American and French physicists and pronounced sound as a bell, yet to date no one has built a device to employ the theory in practice. What more shall we learn, probably of enormous value when properly under-stood and used in practical form, from our visitors in the near future? If time and space potentials are changing, we ought to know as much as possible now.

4

UNIDENTIFIED LIFE-SAVERS

Much UFO evidence from Warminster has been ridiculed and belittled by the so-called 'experts', but I feel elated when so much support from strangers comes my way. A trite saying is that it is easy to be wise after the event, but so much in the accounts tallies precisely with Warminster happenings over the past decade which I had the good fortune – or misfortune? – to report, as part of my news-gathering duties, earlier in the overall saga. No one is ever too wise or worldly to learn.

Here is another minor mystery: a local farmer found, one morning in the summer of 1965, that several acres of land he had left fallow near Warminster were 'a mass of weeds'. They were silvery thistles of a rare type that virtually ceased to flourish in England in the year 1918, proclaimed the experts who rushed excitedly to the scene. The amazing story made headlines in regional newspapers, but we did not dream of connecting this sudden eruption of unusual growth with visitors from space or their machines. Naturalists and botanists from all parts of the United Kingdom flocked to the farm to see the astounding crop of weeds, the unique overnight wonder. They included highly qualified staff from horticultural and agricultural schools and colleges, botanical gardens, universities and natural study groups. They came in awe to view the acres of an extremely scarce plant that only grows in the odd one or two clumps in any county of England.

Although we did not link such freakish growth with UFO activities overhead, it could be that the force field of power from spacecraft had unlooked-for effects on plants of specific kinds. Checking carefully on reports, I discovered later that inexplicable flying objects were regularly seen over this particular piece of farmland on successive nights before the wild thistles mushroomed in prolific and abnormal quantities. Over Harold and Dora Horlock's garden in East Street 'a large red poker' hung suspended for some while then flew off with a noise like the 'crackling of bacon frying'. This story made news copy and a BBC television programme from the West of England, but this was after ordinary thistles in the Horlock's front garden had soared to a prodigious height of almost 12 ft as opposed to the expected 3 ft 6 ins, that same summer. The Warminster area farmer begged reporters to keep secret the precise whereabouts of the remarkable thistle crop, fearing that if it was publicized hordes of the curious would trample down agricultural crops growing in adjacent fields. We respected his wishes. At first, he confided, he was strongly tempted to hoe down the offending weeds in case they spoiled his farm crops, but a naturalist friend prevented such desecration in time, aware of their identity and rarity.

Joel Sibley, a Warminster gardener, grew a plant that broke world records for the number of cucumbers on its stems. A leading horticultural journal seized on this news, with illustrations. It made a splash in the gardening world. Other minor revelations broke that year, not mentioned other than on the journalistic front for the simple reason that their possible connection with UFO activity did not cross my mind.

To bring the amazing Warminster UFO saga almost up-to-date, let us pause to consider the heart-catching experience of a 41-year-old plumber employed by Wiltshire County Council, Stanley J. Curtis, of Alcock Crest, Warminster. He was driving homeward along the Westbury road after a bingo session. It was 10.50 pm on Friday 2 September 1977. The moon was half covered with cloud, but the night was otherwise clear and star-speckled. Stan stopped his car near the notorious tree growth known as Colloway Clump, where his eyes had fastened onto no fewer than four orange-hued aeroforms brightening the clump, transfixed as if glued to the very trees. There were two pairs of UFOs, only a little distance separating them as they hung in lambent splendour, apparently within the clump itself. Then he saw a fifth at ground level, while the remainder careered around to the rear of the tree-belt. The isolated one changed to a silvery shade with a domed top. Two orange circular lights flashed from its upper structure, and at each side of the UFO was a fin that thrust upward, as if it could be folded tight against the body of the object. But it was the peculiar 'apron' encircling the craft at its base which was perhaps the most striking feature; it was fairly wide in depth, segmented by several fluorescent tubes running vertically along the length of the apron, and these were giving off a pale-blue and green light. Mr Curtis told me: 'The fins at either side, similar in shape to those of a basking shark in the sea, looked as though they could be folded upward on to the rims of the air vessel. The orange lights or windows emitted no shafts of light or beams, they were static and glowing. One could liken them to port-holes lit up from inside the craft.' Emotionally overwrought, after four minutes, he sped home 'like the wind' to get away from an unworldly and frightening scene!

Yet this was not Stan's initial encounter with a UFO. Way back in 1951, he recalled, he was walking with a friend up Boreham Hill, when both saw a cigar-shaped aeroform meandering over Southleigh Woods, to the right. And, not long after the Warminster enigma began to spread its puzzling wingless and noiseless forms over the town area, he was working as a plumber with Charles Hudd, mentioned in *The Warminster Mystery*, when a gang of workmen in early morning hours saw a huge cigar over Cop Heap which suddenly, with no murmur of sound, seemed to split in half, five silvery discs tumbling from the opening to scud in differing aerial directions, much to the men's consternation. The matter was reported to Warminster police.

When coming home with a pal at about 10.45 pm on Tuesday 30 August 1977, 18 -year-old Christopher Curtis (Stanley's son) saw an orange-coloured light in Colloway Clump area. Then an object sped very fast towards Cley Hill as young Chris told his parents in their kitchen when he reached the haven of safety at his home. 'Again, there was no sound at all, which is a plain indication that these machines are nothing to do with the military,' the father told me seriously. 'I have never seen anything so mysterious in my life, especially when they brighten colour as they flash away at astounding speed and out of sight in a split-second!' So it was all 'happening' in Warminster area in August and September 1977.

Local investigating body Ufo-Info (49 The Down, Trowbridge, Wiltshire) was swamped with calls concerning not only many sightings of ufos, but actual landings. Regular sky-watcher Bridget Chivers, of Melksham, took photos of the out-of-this world sky giants, one resembling a silvery bowl with a cone-shaped top, another looking like separated halves of a white ball; and the story plus pictures were featured in late August on HTV news (Bristol), in the pages of the *Bath and Wilts Chronicle*, in the Bristol based *Western Daily Press* and *Evening Post*. Another graphic account of landings appeared in the *Wiltshire Times*, Trowbridge, for whom I reported for several years on Warminster and Westbury area news in the late 1950s.

A group of sky-watchers on Cradle Hill, including a police sergeant's daughter, Janine Perry, solicitors' clerk Gillian Ephgrave, ambulance driver Phil Champion, lads from Croydon, London, Cardiff, Hertfordshire, etc., at 10.25 pm on Sunday 28 August 1977, witnessed several satellite-type

36

aeroforms high above our heads; but some changed direction in flight, or hovered, as they sped north to south, so obviously were not man-made objects. But it was when a UFO resembling a shallow silvery bowl shot suddenly from the top of the copse and sped a short way towards where a large group of us were standing, aghast, on the military road nearby, that excitement reached fever pitch among the several dozen watchers. I grabbed a torch and shone a morse code message at it, whereupon it abruptly vanished from sight at near to soil-level. It simply appeared to dematerialize. But Stuart Payne, standing further along the road, ran back to join us, pointing upward. 'There it goes!' he yelled. And we saw it was indeed the same intruder, still glinting in crystal majesty, but now almost starlike in character and speeding along to the south, to all intents and purposes at that juncture like a normal satellite.

Brian Edmonds and Peter Smith, of 38 Maple Road, Penterbane, Cardiff, wrote to me early in September, points from their letter being:

Last Sunday, 28 August 1977, my friend Peter Smith (amateur astronomer) and I had the pleasure of visiting Cradle Hill and experiencing a few sightings in the sky (my very first); and also the exciting incident where one actually rose up skyward out of the copse. I have not stopped marvelling at our good fortune to be there at just the right moment! We had been planning our trip from Cardiff for some weeks beforehand, and were delighted that it all worked out so well, with beautiful sunny weather and a crystal-clear night for skywatching. We left Warminster on Monday 29th, with a little regret at being unable to stay on for a few nights longer.

Since returning home, I have done some private sky-watching from my back garden, now knowing exactly what to look for, which I did not before my Warminster experience. I really didn't expect to see anything over

Cardiff, though. However, on Wednesday night, 31 August, a very clear and starlit night, I saw between 9.45 pm and 10.45, to my great astonishment and wonder, no less than six UFOs at roughly five or ten minute intervals. The first one travelled from west to east. The following two from NW to SE, and the last three were journeying south to north; and I truly believe that, had I stayed on watch all night, I would have seen an endless procession of them.

However, it being a somewhat chilly night, I went to bed still in a state of excitement and awe at what I had seen. It was strange, lying there knowing that 'they' were journeying silently yet watchfully across the night sky – but, by now by no means frightening! Peter and I intend to visit Cradle Hill again in a few weeks time.

A pleasing and reassuring letter from two fine lads who came, perhaps with doubts, but returned home fully convinced as to the authenticity of UFO sightings galore at Warminster since late 1964, some fourteen years ago.

Also in August 1977, I heard that a white and flashing UFO had actually landed on waste ground at the rear of the Yew Tree Inn, Boreham Road, Warminster, close to Boreham Mill, where twelve ley lines are reputed to cross. The landlady of the hotel, Phyllis Butler, awoke at 4 am to spot the garish light, looked out of a back window and was shocked by what she saw near grass-level. UFO-INFO snapped up the astonishing story quickly; and police on motor patrol saw a crimson UFO form that 'split into two rounded portions' shortly before this at Stockton village, near our town. Judging from Stan Curtis's reference to 'two pairs of orange lights' at Colloway, I assessed in retrospect that these belonged to two UFOs, similar in character to the grounded third. The story made headline news in many area news media during early September.

In the nineteenth century, Oliver Wendell Holmes wrote that he thought it not improbable that man, like the grub which prepares a chamber for the winged creature it is to become yet has never seen, may have cosmic destinies he does not understand. Cosmic destinies are long and impressive words, and we have not yet conquered this planet, much less the cosmos. Homo sapiens is only one species of man, there have been many before in our long history, each one a rather better model than its predecessor. The present one keeps on trying to frighten itself to death; it can never succeed. The first fear was black magic and deadly curses of witches, the latest is pollution of earth, seas, skies and minds of men. Black magic is coming round again. It exists but is not supernatural, being a form or state of mind that is invisible and difficult to understand. When nuclear fission became practice, ancient atoms were reduced to quasi-matter long before their due time. Most UFOs – not all – are visible evidence of matter in the process of rebuilding long before the due season. When invisible, they are polluting the air, earth, seas and minds of men. Any volcano in steady discharge, such as Etna, is like a nuclear reactor in operation. An *erupting* volcano is a nuclear fission bomb in essence: the first result is catastrophe; and, later, much increased fertility. One discernible sign of this may be the sudden fertility of once-rare Crown of Thorns starfish, now destroying living coral in the Pacific Ocean area where fission and fusion devices were employed over a period of years. Their fertility is so great that one creature cut into a hundred pieces grows into a hundred more creatures!

In some respects, nuclear progress as engineered by present homo sapiens is dynamite today. The probable existence of another inhabited planet occupying the same space as our own is shown by personal experience throughout the ages. Today, spiritualism has become quite respectable. Reincarnation is a puzzle rather than a gimmick. Ghosts have been photographed. So have UFOS. Books and records describe how inexplicable sounds of tape recordings were slowed, then raised in pitch, to reveal the presumed voices of dead homo sapiens. In an inhabited world of spiral space time, this speeding up of solar time is inevitable; and with some intelligent mathematical treatment by open-minded investigators the myth of the so-called supernatural should soon be banished. Nothing rises above nature!

Kevin Murphy, a young journalist working in London, is the only amateur to swim the English Channel both ways non-stop, and has had many hair-raising adventures. 'I have escaped from tricky situations too many times for it to be just a coincidence,' he admits. On one occasion he was swimming in Weymouth bay in Dorset, popular seaside resort in the west country, when a heavy storm blew up. It was so severe that, to avoid being swamped by huge waves, the pilot-cum-safety boat had to circle him instead of staying alongside as usual. Suddenly, Kevin realized that the boat had not circled him for some while. He was not unduly alarmed, simply puzzled. He turned round and saw it way back in the distance. A lobster pot had fouled its propeller. By the time that the crew of the vessel had cleared the obstruction they had lost sight of the swimmer in the surging seas and lashing rain. Meanwhile, Kevin was being drawn by the savage tide towards some dangerous rocks about a mile-and-a-half away. After twenty minutes of frantic searching, the boat crew were on the verge of calling out the lifeboat, the emergency warranting this extra assistance. But suddenly an inflatable power boat appeared on the scene as though from nowhere. In it were three men dressed in black. They spotted Kevin and beckoned the pilot boat to come over to him, then they simply disappeared in the driving rain as mysteriously as they had arrived! One moment they were carrying out their good Samaritan act in the height of the storm, conspicuous by their dark clothing – the next, they were gone as though swallowed up by an out-sized whale. 'It may seem just a coincidence,' says Kevin, 'but the appearance of a small dinghy in the middle of a bay during a storm, apparently not heading anywhere in particular, is – to say the least – unusual.

'I remember at the time I was praying like mad for help; then the dinghy appeared. And when it turned up, even the crew of the pilot boat felt it could have been a divine apparition. There was no sign of the rescuers in the choppy seas all around us.' Kevin, who has had several other escapes from dangerous situations and predicaments, including a car crash from which he and his mother emerged without serious injury, says simply: 'It all adds up to God's protection'. In the story is no reference to UFOs, only a thinly veiled allusion to a Great Intelligence; but to me the story makes a welcome and wholesome change from all those lurid nonsenses about hostile entities from Outer Space and alleged abductions of aeroplanes and sailing craft from the notorious currents of the Bermuda Triangle.

These are but a few of thousands of reports I have handled and evaluated from all sectors of Britain alone, in the past fourteen years. At first they make refreshing and stimulating reading, but the novelty wears off when they become too numerous, almost to the stage of monotony, which genuine UFOs should never be! We have been presented with a vast assortment of preferred alternatives to UFOs by government departments and ministries in the past, for example the much maligned planet Venus, sun dogs, refractions and reflections, spots before the eyes, optical illusions, mental delusions, hallucinations, ice crystals in the atmosphere, lenticular clouds, and so on. Thus, after the public has been so efficiently brain-washed, the shock of seeing an inexplicable aeroform that 'just does not exist' has a severe and traumatic effect on some witnesses, which is psychologically shameful!

What can the honest inquirer believe when confronted by

two contradictory statements inspired by a government agency? A sighting in the UFO 'flap' of 1973 was made at literally the highest level. Associated Press carried this report on 18 October:

> The Skylab 2 astronauts in debriefing sessions told of seeing a mysterious reddish object in space, the space agency reported yesterday. The disclosure came amid a new nationwide rash of sightings of unexplained aerial objects.
> Dr Owen K. Garriot, one of the astronauts, said the object was brighter than any of the planets, had a reddish hue and was not more than thirty to fifty nautical miles from the spacecraft. The astronauts saw it one day in mid-September, but never again in the 59-day flight.'

However, another news report carried this somewhat contrary statement: 'A spokesman for the National Aeronautical and Space Administration said that crews of Skylab 1 and Skylab 2 saw numerous UFOs while in space.'

Although mentions of flying saucers and possibilities of life elsewhere in our solar system evoke natural scepticism from professional astronomical and biological fraternities alike, so many science fiction predictions of the past have become accepted realities today that visual experiences of millions of testifiers to glowing UFO shapes in our skies can no longer be sensibly disparaged and ignored.

Too readily perhaps, we assume that all forms of humanoid life must necessarily conform to carbon-based molecular structure, with silicon or multicelled types of animated intelligence out of the question! Despite voluminous amounts of evidence from rational people throughout recorded history which points to the probability of ghosts, spectral shades, phantoms, poltergeists and sub-astral entities definitely existing and manifesting, the official attitude seems to be that all must be attributed to hallucinatory incidents bordering on the supernatural; otherwise, there would be a situation tantamount to open admission that other-dimensional kinds of life may belong to beings far superior to mankind on Earth. Basically, might that constitute the frightening prospect which induces most of our governments to clam up and (conversely) clamp down on 'this infuriating and never-ending UFO business?'

There arose a rumour in Warminster area just after the 'slithering serpents of sound' described in our first chapter. Vainly I tried, but could get no official confirmation of an 'impossible' news story told to me quite seriously by a number of soldiers and civilian workers at the School of Infantry. Briefly, it concerned newly-installed and foolproof electronic apparatus used for raising and lowering targets on a rifle range not too far from Battlesbury and Starr Hill. Of Swedish design and specifically with NATO Forces' firing drill in mind at the military buttes, the officer or NCO in charge of soldiers' shooting simply pressed a button that controlled elevation and depression of cumbersome targets on the range, thus obviating manual and time wasting hauling up and pulling down of the targets by pulley-rope methods in operation during my own serving days in the Grenadier Guards. If I dared credit the soldiers' stories as true, the 'foolproof' system broke down during the chaotic fortnight of the Starr Hill slithering snake sounds of subterranean nature. The targets rose and fell in a haphazard unpredictable way, even though the man in charge dutifully pressed and released those vital buttons to secure correct firing returns allocated to the wretched marksmen. Targets rose, sometimes en masse, sometimes singly or in odd groups, and lowered at intervals even when the electronic 'control' was manned! It was as if an ominous underground force was making the targets behave in a most erratic manner, out of the control of any human agency and 'taken over' according to the whim and will of an unseen power.

Observant readers of the flying saucer information broadsheet *Ufo-Info* noted a peculiar news story that broke on Bristol-based TV networks concerning night shift employees at Meare Down Quarry in Somerset (on a ley line with and not far distant from Warminster), where strange orange lights plagued the quarry and were seen by reliable workers, setting darkness ablaze with roseate glows. They manifested on several occasions and forklift trucks were taken over by an uncanny force, with controls torn from the petrified operators' hands!

Assuming there is truth and no falsity in accounts furnished me by School of Infantry troops and civilian staff at Warminster, also the weird experiences of the fear-filled people at the nearby quarry, is there any connection or significance between UFO jaunts at Warminster for the past fourteen years and these adjacent mysteries?

That elusive property we term Time should present us with satisfactory (or unsavoury?) answers in due season. As an investigator at Warminster since 1964, I can state that UFOs take the varicoloured shapes of glowing spheres, luminous orange peardrops, ovular jewels that range from blood-red rubies to winking diamonds of flashing light; from lustrous grey daytime pearls with a shimmering surface to fiery green emeralds that decorate the nocturnal heavens with fluorescent brilliance; transparent opals to dense white cigar aeroforms. Perhaps differing colours are employed deliberately to appeal to and enhance the mental and spiritual evolutionary progress of each individual viewer?

5

COPPER MOONS

There are cases recorded where house lights have flickered synchronously with pulsating UFOs, and others where people suffered electric shock, minor burns and irritations as well as headaches due to highly charged surrounding areas when UFOs are very close. The electromagnetic effect also offers a solution to the mystery of small craters associated with landings of the craft in America and England. This could be due to the approach of two bodies with different potentials and polarities, and the resulting electrical discharge. As we know, the famous unified field theory of Einstein indicates that electricity, magnetism and gravity are all manifestations of one force. An artificially created gravitational field, by means of electromagnetism, can explain the effects associated with UFOs, including the silence.

For instance, the G-field explanation accounts for the reason why the spacecraft can withstand the friction which would normally be caused by such fantastic speeds through the atmosphere that have been observed visually, and simultaneously checked by radar. We know from our physics that if an object moves rapidly though molecules of air, the friction causes a positive charge to be formed on the surface of the object. By an elementary law of electricity, we know too that like poles repel and unlike poles attract. Thus, by inducing a positive charge within the machine when it is moving rapidly through the atmosphere, the molecules would be repelled, tending to produce a narrow band of vacuum around the hull. This would reduce the friction effect almost to nil. And by a simple law of sound, we know if there was any noise associated with the object, the sound would not pass through the vacuum, although a low humming noise is sometimes heard when UFOs are low, and moving slowly or hovering. Additionally, a G-field would drag surrounding air along, so there would be no turbulence and this factor would further reduce friction and account for the silence.

Sometimes, to determine what the future holds in store for man, one has to survey not only present trends but the patches of light and shadow cast by the immediate past. In Ufology, where the brush of history may paint vivid colours in the pages of years to come, it is clear that pointers in the past can be highly relevant to the course of future events. Take, for example, the flaring of flying crosses in the sky. They are nothing new in aerial flights over the notorious Warminster area, although there was a spate of such reports in late October of 1967, given wide and fair publicity in the British press at the time. Reliable witnesses included police officers, senior RAF officers and army personnel. To recount all reports would be boring, tedious and unnecessary in the contents of a book aimed at raising the thoughts of the enthusiast to loftier horizons than a narrow skyline of mere UFO sightings that have become so numerous over recent years.

Observant readers will recall similar testimony given by people in *The Warminster Mystery*. The terminology then was 'four-pointed stars', that moved at varying speeds and trajectories, slicing through puffs of cloud as a knife shreds culinary vegetables. They were especially active in September of 1965, as the book stressed. Sometimes they were likened to 'flying swords' and one witness even referred to this singular UFO type as an Excalibur, legendary sword of King Arthur of Round Table renown. When low in altitude, one can see around the shining outline of the cross itself a more subdued amber casing in circular mode. Thus, they conform to the usual UFO rounded contours in overall shape, character and dimension. For the four-armed cross to stand out clearly and outshine the outer casing, the spaceship must be viewed at extremely low height.

According to scientific concepts in engineering design of conventional aircraft, these glowing chariots in space are aerodynamically unsound and utterly wrong in flying principles. If only because of this reason, they have no right to use our air space, claim the obstinate and those wishing to adhere to dogma, spurning the serious testimony of reputable witnesses who are their equals in realism. Nearly all the experts, I recall, lambasted the delta-winged bomber when at the designing board stage years ago. Yet it flew, and very ably as it happened, so ramming the ultra-conservative views of these same experts down their throats! The humble bumble bee, with its massive body and relatively tiny wings, can still teach us lessons.

Recorded in many historical documents apart from the Bible, Talmud, Koran and Sanskrit writings, becoming more prolific in number since the last war in all major countries of the world, UFOs are factual to a large proportion of a no-longer-gullible public. Sensibly, these people refuse to be fobbed off with weak official explanations that fall dismally short of truth. Two policemen had an 80-90 mph chase after what they described as an unidentified flying object over a Devon road. A spokesman for Devon and Cornwall police told reporters it was very large, bright and in the shape of a cross, according to the constables. One news medium claimed the officers later reported seeing the aeroform again, when it was this time joined by a second similar 'thing'. PCs Roger Willey and Clifford Waycott were in a patrol car at 4 am when they first spotted the glowing object, on the Holsworthy to Hatherleigh road. PC Willey is based at Okehampton, PC Waycott at nearby Winkleigh. Both are married and, as indeed are all police, keen observers. The nearest they got to the air chariot, they attested, was forty yards.

47

It was then 'at about treetop height'. The spokesman added: 'It followed a course virtually over the road on which they were travelling and they were doing 80 mph at one time; then it left them. It appeared to stop in a field. They stopped also and got out – but it disappeared.'

About the time that the two Devon policemen were engaged in their hectic pursuit of an aerial chariot, Lieutenant Michael Casey, on military exercise near Imber, the deserted ghost village on Salisbury Plain near Warminster, saw something of similar nature in the Bulford direction. He was based at the School of Infantry, Warminster, serving with the 1st Queen's Regiment at Lingfield. He says:

I have got an open mind regarding what it was. At the time I thought it was a satellite or a bright star. But when I saw the television news, I realised that what I had seen was very much on a par with the Devon police sightings. So I reported it to Warminster police. I was on exercise about one and a half miles south of Imber at 4.30 am. We were on a two man patrol, but I did not mention it to my partner because he was busy map reading. It was about thirty degrees above the horizon. It was a white blob with four prongs, roughly twenty times brighter than the North star and ten times bigger. It appeared to be stationary and I watched it for some fifteen seconds, although it may actually have been there longer. There was no sound and what I saw was apparently much farther away than what the policeman saw. I have done some basic astronomy and it certainly was not Venus, as somebody suggested after the police sighting.

That was the report made locally by the officer – and it is but one of a number that Warminster police receive each year. We are extra careful in our meticulous assessment to define true UFOs, yet I always welcome descriptions given in this frank and excellent way. Concise, to the point and breathing very truth. No journalist can ask for, nor expect, anything more in a sometimes bewildering quest for the factual where UFOs are concerned.

Our trio had actually logged definite UFO landings on our sightings sheets by then. In fact, at Starr Hill on the night of Thursday 13 April 1967, we not only saw the glowing hulk of a UFO at a range of 250 yards, weaving and rolling over the shoulder of the hill, as large as a house window at thirty paces; but also observed a wonderful display of conical lights that flashed from its revolving rim as it practically sat on the ground near a farmhouse a thousand yards away. Later, after the giant craft mysteriously blacked-out in mid-flight, we heard peculiar coughing sounds coming from the lower slopes of Starr Hill and the unnerving squelching of jackboots or wellingtons in soggy soil and soft mud around the barn near our parked car. Bob Strong and Sybil Champion were my companions on that mind-searching night. We told a few people about our experience and how we kept watch for several hours until – quite frankly – the gurgling mud-sucking footsteps of an invisible walker stretched our nerves to breaking-point. They could stand no more of the tremendous pressure, and we hurriedly left an eerie scene at about two o'clock in the morning.

We told very few people, for who would credit we were not simply shooting a line and fabricating such an astounding sequence of events? One so-called paraphysical expert from Salisbury openly derided us and contended, after Bob solemnly told him the full story three times, that what we had seen was a farm tractor, with its lights on. This was an incredibly naive accusation! Between 11.45 pm and 2 am, would a flying tractor have been operating with three balls of light that sped from left to right and returned at lightning speed to a blacked-out craft, dizzying to watch? When true narratives of personal experience were laughed at and scorned by obtuse 'experts', we withdrew and kept certain information strictly to ourselves. Remember, we had nothing to gain from deliberate falsification of evidence, and everything to lose – including our good names and jobs.

Michael Wakely, chairman of Wiltshire UFO Research Group says:

> Reports of unidentified flying objects are still being received by research groups up and down the country. Warminster is still an active area. The object of these groups is to investigate fully these phenomena. This is something the authorities will not or cannot do! The public showed that they were not satisfied with investigations into the Flying Crosses seen over Devon and elsewhere. Two main explanations were given by the Ministry: (1) That the lights were the planet Venus. This was emphatically denied in a great many cases. No police officer is likely to go around chasing planets at eighty miles an hour! Is any trained observer, such as a police driver, likely to be so simple as to give chase to a planet? In many instances, Venus had not even risen by the time the crosses were seen. (2) That the crosses were aircraft refuelling. This was absolute rubbish! Major Westgrove, director of the USAF base at Ruislip, denied that any of his planes could have been mistaken for crosses because (a) red lights were very prominent (only white crosses were noted) and (b) all exercises had finished by 9 pm (most of the crosses were seen in the early morning). Until a Sunday newspaper mentioned this, the Ministry of Defence had stuck to these weak explanations; then the following announcement was made: 'It looks as if there is still no rational explanation for the objects the policemen report having seen.' All I can say is: they did not try very hard to find one. They just buried their heads in the sand in the hope that the crosses would soon go away. If we had another 'saucer' flap tomorrow, we would be given the same inconclusive explanations that even children of nine or ten would refuse to accept! They are insults to the intelligence of the thinking public. Research into UFOS is being carried out by

major powers all over the world. The USSR, under General Anatoli Stolyerov, have formed a panel of eighteen top Soviet scientists to investigate the problem. U-Thant stated publicly that: 'Next to Vietnam, UFOs are the world's next big problem.' UFOs have been seen in their thousands since 1947. Not every light in the sky is extra-terrestrial, and most of them can be explained away as hallucinations, sundogs, refractions through ice crystals, etc., but not all. People have a right to know what is going on in the skies above us, and we are trying to find out as much as we possibly can.

Even though the UFO background at Warminster is in its special way unique, with so many prototypes and precedents set there so far as aerial perambulations of these flying mysteries are concerned, it does give one cause for deep thought when similar outlandish things are reported from other sectors of our planet, perhaps years later. Intuitively, one can spot the genuine from the fake, or the UFO which is a hoax or practical joke. Yet I must honestly challenge scientist or layman alike to produce positive proof that UFOs and their intelligence are absolutely identified. UFOs or unidentified flying objects is still the correct way in which to describe them. We can hazard pretty shrewd guesses, depending on the gamut and range of personal sightings, plus appraisal of weird on-the-ground incidents that are certainly not of earthly origin. We cannot afford to be dogmatic! There is room and space for varying shades of opinion. A golden rule for any journalist, especially if his shorthand is patchy or his data insufficient, is: 'When in doubt, leave out.' The same applies to research into UFOs.

I recall coming off Cradle Hill with Bob Strong one morning just before dawn. When we had motored down Elm Hill and reached a new housing estate in Portway, Bob pulled the car to a halt as we both gazed at what looked like a huge coppery ball of light, several times larger than the moon, on the western horizon. 'It is certainly not the sun-rising – that will be in the east,' I remarked. Bob looked at me seriously, his brown eyes gleaming at uncovering a new mystery. 'It isn't the sun, Arthur, nor is it the moon. That came up hours ago, remember? In the opposite part of the sky, too.' The ball shimmered and gradually sank behind the playing fields of the new council estate. It was a further puzzle to us. But echoes of it tiptoed through my mind when I heard of a visual experience of a physiotherapist at Sudbury Hospital in Suffolk. Here is her unembroidered story of 'an enormous copper ball, as big as four moons, following very low the course of the river through Sudbury, and travelling from the Colchester direction through Long Melford.' It is told by Dorothy S. Hawkins, of Carew Road, Eastbourne, Sussex:

I was then physiotherapist at Sudbury Hospital. I remember it journeyed at walking speed. I left the glass balcony ward with four men in it, and on my way home at 7.40 pm I saw this extraordinary thing through an opening between a public house and a butcher's shop. My urge was to rush into the opening, as I could have gone home that way, but it was private property so I ran along the path and into our garden – but the thing was gone. Next morning I went back to the same ward and remarked that I had seen a strange thing on my way home. The four patients at once said: 'Not a big moon? We saw it pass here!'

Their windows looked out over the River Cornard side of the town. I saw the object on the Melford side. A few minutes later one of the men, reading a daily paper, said: 'Here's your moon. They call it a meteor.' He cut out the piece and gave it to me. I still have it somewhere at home. I left the flat I was then living in in 1936, so it was before that. I had never heard of UFOs until 1961. I have seen several lights in the sky but I did not know them by that name, I just called them 'hovering stars'. One night about eight years ago at Sudbury, I looked up as I was getting ready for bed. Something was nagging at me to go out of the back door into the garden, so I finally went and I looked up – and there was one just over my head, travelling due south. I called out: 'Thank you.' I shot it a telepathic message: 'Please twinkle, or do something to let me know I got your message and that I have seen you.' The UFO promptly zigzagged sideways three times, then proceeded as before. I am no judge of speed, but it was slower than an ordinary aeroplane and much lower. There was no sound.

Flying chariots are pretty real to those who see them, even if government departments pooh-pooh their existence. Take a typical case involving a very down-to-earth Royal Marine commando and what happened to him at one o'clock in the morning on Bank Holiday Monday in August 1975. He was so surprised and shaken that he reported the puzzling incident to Warminster police. What is more, he had never heard of the infamous Warminster Thing! Nor had he been drinking, a point that was stressed to me by PC Bernard Hollands at the local police station. But his nightmare experience will live vividly in his mind for ever.

Stationed with his unit at Arbroath, RM Commando Andrew Ian Simpson, aged 22, was travelling by car near Lords Hill, three miles south of Warminster, having been driving for about an hour *en route* to Devonport. He suddenly noted that his car was being hotly pursued by a bright-red globe of light, which appeared to come from a field on the offside of the highway. The gleaming sphere rose and sped alongside his car, bobbing up and down. The headlights of the vehicle went out, the sidelights flickered, and the engine began to misfire. This continued for several seconds, he told police, until the spherical aeroform floated away to vanish in a downward direction. There was not a trace of sound, which he found the oddest feature of all, for the UFO globe was so close to the car. 'He was extremely unnerved and severely shaken up,' PC Hollands told me. 'He was a deathly white and blurted out his story excitedly yet clearly. He afterwards lapsed into a bewildered silence. He had definitely not been drinking and there was utter sincerity in the way he told a weird tale.'

The police said he was by no means the first motorist to report these 'things' flying along lonely roads around Warminster area. Two army majors had had such 'baptisms' from unexplained aerial vibrations like heterodynes or refrigeration units of thousands of horse-power. Army personnel had reported 'spinning tops of fire in the sky' and globes of light sometimes swooping down to car-level. Over a hundred people gathered at Starr Hill on the same bank holiday Saturday in August 1975. Several flying objects were seen in a hectic hour. Some were not silver-white and regular in course, as satellites should be when reflected in the rays of the sun, and all were noiseless. On the hill were police, soldiers, engineers, amateur astronomers, students, a computer operator, two professional photographers, police families, a County Council official and many others, coming from sundry parts of England including Surrey, Durham and Essex.

They saw a fairly large UFO, glowing pinkish-red in colour, that erupted over a burial barrow on the hill line, then bobbed along a few degrees above the horizon like a duck on rippling water. It vanished when car headlights struck it, from the road, but not before professional photographer Chris Waller, of Upper Marsh Road, Warminster, captured its image on film. All witnesses were impressed by the lack of noise during the dancing display from whatever it was, and Mr Waller said: 'I really cannot put a name or an identity to it. It was certainly no star or planet, and it was much too low for a satellite. Besides, the colour was wrong for that. I have never seen anything like it before, but this was my first sky-watch, anyway. Veteran observers on the hillside were convinced it was something from another world. It beats me what it could have been!'

6

UNKNOWN VISITORS
LAND

It is doubtful whether Chris Trubridge or I will ever forget a certain Thursday evening in June 1977, and the astonishing series of incidents on the Warminster UFO front that we were privileged to witness at close range. Suffice to state that the overall experience made a lasting impression on us because it was so unexpected that it took our shocked senses several weeks to recover from the sheer loveliness of it all.

It was 9.25 pm when we walked together up Elm Hill *en route* to a proposed sky-watch session from Cradle Hill. The sky was a brilliant expanse of blue, speckled only by a few scattered white flecks of cloud in the distance. It looked all set for a lengthy period of clear, dry weather ahead, so that raincoats were the last things on our minds as we trudged slowly up that first of two fairly steep climbs towards our destination. We had reached a point opposite the road winding leftward to the West Wiltshire Golf Club, and stopped for a while to regain our breath and to lean over the gateway by a stile giving direct access to the prominent cluster of trees to our right known as Copheap, where tradition has it that an early Saxon chieftain and his family were interred in the bald patch of earth on the top of the mound in the midst of tree growth.

Almost instantly, we were aware of a bright star-like object high above and in front of us that was moving very slowly from the direction of the ancient earthwork and Iron

Age fortification of Battlesbury. Then it increased rapidly in speed and expanded in size as it abruptly zoomed in to where we were standing in rapt silence, dumbstruck by the magic in motion in miniature. Not a word, only a sigh, passed between us as our eyes followed this daring phantom of the firmament.

The shining spheroid slowed and stopped near us, then changed from its starry appearance into the more defined form of a crystal ball or saucer-shaped aeroform. Its outlined edges and rims were hazy and fluctuating as it loomed noiselessly nearer and came lower than the topmost tip of the conical tree-lined Copheap.

As it remained static some twelve to fifteen feet above ground level, glowing with its silvery light that dazzled us by its brightness, we could clearly distinguish that, jutting upward from the semi-spherical 'body' was a slender crystal-white pole or aerial, with a round crimson ball-like appendage at its terminus. It was rather incongruous atop the UFO, as it swayed slightly on the wavering supporting pole, the latter seeming to suffer from a shivering fit, but both pieces of joined apparatus constituted an integral part of the whole gleaming structure of the vessel in the air. We stood, mute and riveted to the spot by the brilliant glare of light that suffused and surrounded the UFO. Maybe it was the dazzling effect on our eyes, but the wheel from the heavens seemed to be spinning or swirling slightly, its rims plainly evincing this perpetual movement from left to right, although the main body could have been stationary. It all happened so quickly and unexpectedly that we could not be sure when discussing the aerial vision later at Chris's home.

As the upper bar of the gate bit into our eagerly tensed arms, the scene changed before us. In a motion that checked the spinning of the lower section, the UFO shifted smoothly to our right and did a complete detour of the tree-clad slope. We watched in stupefied silence as we clung white-knuckled to the topmost stave of the gate, to see it vanish from sight behind Copheap in its sudden circling manoeuvre. At this stage it was a silvery ball of light that hovered momentarily by the tops of the trees, then slid effortlessly down the limbs of their trunks to shape an indistinct blob of lessened brightness halfway up and at the far side of the memorial to the dead of two world wars.

We were anxious, impatient even, for it to reappear, but it sauntered very slowly along behind the trees. The UFO had outgrown its size and had a distorted shape, due to its semi-concealment by the trees standing now like silvery sentinels between watchers and apparition; but we could still discern that strange crimson ball, oddly flickering at the tip of the crystal aerial, as it skirted the outer fringe of trees and hid from our gaze in its painfully sluggish movement. It was one of those rare occasions when Chris had neglected to bring his camera along, which nettled him, but I carefully noted all details in a notebook with a stub of pencil, a habit of my news-reporting training.

Expectantly, and seething inwardly with excitement, we waited for the celestial chariot to reappear, confident it would leave the rear of the tree-belt and emerge in seconds. But it failed to do so, straining our patience to the limit, although we stayed clutching the wooden support of the gate for several minutes instead of jumping over it to pursue the alien quarry that bathed the whole area with silvery streaks and fragmented pools of light.

With one accord, Chris and I decided to 'give up the ghost' and moved along the road to another gateway some sixty yards away at the junction of land at Parsonage Farm and a road leading to married quarters of military families. From our new observation post we scanned Copheap eagerly, and then, still gliding slowly at well below human walking pace, we saw the glinting splendour of the silvery giant as it swept in undulating fashion from behind the rim of the foliage-covered mound of Copheap.

It hovered there for some time, possibly half a minute,

57

then, suddenly and dramatically, it sped at lightning speed upward into the blue sky with its cottony tendrils of thin cloud; and seemed not only to shrink in size but change its whole shape and character. Incredible as it may appear to non-believers, it now had the identical shape and navigational lighting of a conventional aircraft, and even simulated the throbbing sound of a military plane as it casually sauntered over the barracks of the nearby School of Infantry! It put to shame any trickery performed by any professional magician in an amusement hall, such as producing rabbits from empty hats! 'Impossible!' you may say. But that is exactly what happened, and we were both staggered at the revelation our eyes and minds beheld and boggled at...

Watching the 'plane' fly so casually eastward, no longer in metallic brilliance, or with its red-glowing orb mounted from a silvery rod-like aerial over its superstructure, we gasped and almost refused to accept the visual fact presented to our staggered senses. Then, in the wake of a truly astounding series of unforgettable events, masses of darkly ominous clouds loomed overhead from nowhere, it seemed, appearing in the sky from all directions. We were standing near the farm itself, debating whether or not to continue to Cradle in view of a heavy downpour of rain that cooled our sky-watching ardour somewhat, when John Rowston (of *Ufo-Info*) drove up in his car from the hill direction. He had been watching from the noted copse at Cradle, and had seen the 'plane' fly over the army zone, earlier, when it confronted us and confounded our senses.

Another true story concerning our flying friends comes to mind, as it happened in 1976 not long after the latter incident. From a reliable source employed at the School of Infantry, which I readily gave in late July to a CID detective inspector and a detective sergeant from the special UFO branch (one from Manchester, one from Brighton) to help them prepare a thesis on unidentified flying objects, during which the senior detective revealed that he had seen a definite UFO in the

Lancashire area; surprising news filters my way. I learned that an Army film unit were screening the taking-off of man made rocket-type guided missiles of ground-to-air nature from low-loading transporters on Salisbury Plain. Their film picked out the trajectory and flight paths of the missiles (probably Honest John or Blue Streak varieties), watched by an assemblage of troops and civilian staff members. The information reached me from a worker at the administration wing of the unit in our area. All went well and uneventfully until a startling sight was captured faultlessly by the camera. A glistening torpedo-shaped UFO peeped from a cloud and swished noiselessly downward, in range of the cameras. It then completed three circular motions around film crew and troops, slowly and deliberately, before zooming upward at a terrific rate, to disappear once more into the cloud layer. It was golden in colour and triangular when seen head-on. The whirling aeroform emphatically flew low over heavy tanks taking part in the staged display before it finally plunged into the curtain of cloud. I tried to secure a copy of this particular piece of film showing a UFO in eye-popping action on an unexpected aerial spree, but so far as I know the War Office possesses the prized film strip and keeps it 'under wraps' for defence security reasons. I am amply assured this is a true story, a number of soldiers vouching for it as factual. (They saw the processed film version as well as witnessing the original happening.)

In February 1978 these mystifying UFOs, elegant orange or gold air chariots that bemuse millions of privileged spotters of the ineffable unknown in our atmosphere over the past thirteen or fourteen years at Warminster, were continuing to haunt local air space. If their intent was hostile or the least bit unfriendly, they could have smashed our defensive systems to smithereens by now and made captive every man, woman and child on planet Earth, as they could have done (past evidence of millennia proves this possibility) at any period of man's history extending back three million years. Why have they not done so? Because they are divinely inspired; are Wheels of Heaven that have always been our custodians and guardians against dark forces of evil unleashed whenever the light of truth and goodness dares to enlighten hearts of men everywhere . . .

Here is the simple and unvarnished tale of four young VIPs (very innocent pals) who were shocked when playing soccer in a field near their homes in Imber-road, Warminster, on Sunday afternoon of 5 February 1978. As the leather ball shot upward from an erratic kick, four pairs of eyes fastened in amazement on to a magnificent UFO which hovered in silver-grey brilliance above Copheap, the ancient early Iron Age conical wooded hill, not two hundred feet from the playing pitch. The boys were two sets of brothers, Gerald (12) and Christopher (10) O'Connor, Nicky (10) and Kevin (8) Salmons, of 76 and 70 Imber Road respectively. Their experience was featured on Harlech television, Bristol, shortly after it took place. Highlight of the feature came when the TV interviewer asked a large group of New Close School pupils whether they believed in UFOs. The deafening response was a chorus of 'Yes!'. This was one of the most recent sightings at the most UFO-haunted town in the world and the latest of thousands of flying saucer manifestations in the district over the past fourteen years. The youngsters, all keen skateboard and fishing fans as well as football fanatics, watched enthralled as the strange aeroform dazzled their vision for three minutes at 3 pm on a sunny day heightened by a pure blue sky. They told me:

It really was huge; it was like two rounded plates stuck together with a wide ring or band of lighter grey separating them. When it tilted, before flying into a puffy white cloud and disappearing, we could see a bluish-green pulsating at the base of the bottom part, dead in the centre. We heard a high-pitched humming note that died away as soon as the Thing went out of vision as if swallowed up by the cloud. But until it tilted and turned, after hovering, it was as quiet as the grave! No noise at all. It was out of this world and made us feel all funny inside! We all rushed off to Warminster police station to tell what we had seen. They advised us to see Mr Shuttlewood. We were so pleased he did not laugh at us, but took our true story seriously. And hardly one of our mates at school laughed at us or took the micky, either. It was so bright! It stood out clearly in the blue sky above the trees. I thought it was going to land on our field, it was so close. To tell the truth, we were all scared to begin with – but we just *had* to report it in case it had frightened some of the older people in the area. When we told our school pals and teachers, they believed us, thank goodness; and most of them backed us to the hilt when we said that UFOS do exist at Warminster. We thought it best to go to the police immediately. The thing spoilt our game of football, any-way!

Perhaps it did not approve of the boys playing soccer on Sunday? The quartet agreed that, when first spotted as their football soared aloft, the UFO made not a whisper of sound. The distinctive humming noise was emitted as it tilted and spun before vanishing into the cloud. It did not emerge, and the noise petered out. The wide band of lighter shade sep-arating the two 'halves' of the saucer seemed narrower at one end, where the two portions appeared to join; or as Gerald O'Connor conceded: 'That could have been due to our angle of vision and perspective at the time. Chris and Nicky have seen funny orange glows in the night sky around here, in the past; but the close-up on Sunday in daytime was outstanding! If it had not been for our football, we would not have seen the enormous high-flyer all silvery and shining above the hill, where a Saxon chieftain is supposed to be buried,' he added.

Warminster police affirmed that many leading citizens, senior Army officers and others have witnessed bizarre aeroforms since Christmas Day 1964, when the phenomena were 'born' in a series of loud bangs and sundry frightening noises at rooftop level. Two police sergeants and several constables (and their families) have reported weird lighted shapes in night skies, too. I think that the 'wheels' will doubtless spin many times in 1978 and future years. Of that probability, we are sanguine.

Peeps at past UFO aerial sprees in the Warminster area are as exciting, and valid, as those of present times in 1978. Retired Wing Commander and fighter pilot Andrew Deytrikh, of Wokingham, Berkshire, had an eye-dazzling experience in Warminster region at 8.35 pm on 3 April 1969:

My wife and youngest son were with me about two miles along the Wylye-Chicklade-Mere road between 2030 and 2100 hours. The conditions for night viewing were excellent; no mist, no clouds. We had not been in position for more than five minutes and, whilst scanning high and to the south, I observed quite suddenly the appearance of what I would have said was a meteorite travelling from SE to NW in a straight line. We observed it for a period of twenty seconds. In my estimation it was well below two thousand feet, was quite silent, and the intense brilliant white light conformed to the configuration of a tadpole or peardrop shape. There appeared to be a rim of orange on the forward rounded end, while the tail seemed to be of a whitish blue. It was a magnificent sight! The light disappeared as suddenly as it had appeared over Warminster, without seeming to slow down in any way. My immediate reaction to the 'meteorite' was: Why no noise at such a low level? There should at least have been a sonic boom, but not even the sound of rushing air, which was very strange indeed! Normally one sees meteorites burn out at very high altitudes, but if one freak should get through then the tell-tale light of the meteorites as it continues its rapid path through the thick layers of the atmosphere should be with it from about seventy thousand feet down to whatever height it burns completely out; and therefore it can be seen for a considerable time. It would not all of a sudden appear, especially if an observer happens to be scanning that particular part of the sky. He would have picked it up at some considerable distance away.

The retired RAF officer and family saw other peculiar things in celestial reaches over the Easter period. We are all believers in practical research at the scene of the inexplicable as a prerequisite to any developing hypotheses. Confirmation of this unusal sighting came from nine other independent witnesses.

Unexpected 'red carpet' treatment was given by UFO intelligence to Robert Chapman, science expert of the *Sunday Express*, when he came to Warminster to compile material for his book *UFOs*. It was a dismally wet night, heavy rain showers lessening the possibility of his seeing our 'visitors' under such palpably poor viewing conditions. Yet he *did*! Dr John Cleary-Baker, editor of *Bufora Journal*, the official organ of BUFORA, reviewed the book and Chapman's visual experiences of inexplicable aeroforms over Copheap, Warminster, in a hushed nocturnal hour. In his review he wrote:

> The author went to Warminster and visited Cradle Hill in the company of Arthur Shuttlewood. While there, he witnessed aerial manifestations which he admits himself unable to explain away. One more testimony to tip the scales against those who believe that the Warminster Thing is a phantom conjured up by Mr Shuttlewood and your editor, in unholy alliance with the little town's Chamber of Commerce. In view of his position as science correspondent of a leading British Sunday newspaper, it it interesting and possibly significant that Mr Chapman's

terminal chapter deals with the recent Condon Report on UFOs in what can only be described as a very restrained and unenthusiastic manner!

Thirteen people were in our sky-watching party at Cradle Hill on the evening of 2 August 1969. An unlucky number? Not so far as a most dramatic double UFO sighting and landing are concerned on what one can term a truly auspicious occasion. The following factual account can be accepted or discredited by readers – that is always your prerogative where the rare subject of Ufology is involved. When in honest doubt, trust your own intuition; but thirteen witnesses is pretty substantial backing as I know from my own life as a news reporter.

My team-mates, Bob Strong and Sybil Champion, left the hill at 9.30 pm after we had been observing for about an hour without positive results. They went to Starr Hill, midway between Battlesbury and Scratchbury, other nearby and well-known viewing peaks. The remaining eleven, many meeting for the first time, were: an ex-naval commander's wife, Mrs Kathleen Bent; her friend, Mrs Eileen Keck, formerly of Winchester and now in Claremont, South Africa; Ian Cowan and his wife Kathryn, from Bournemouth; American Gwen Smith from Seattle, USA; Christopher Trubridge, of Warminster; his musical student friend Robert Coates, from Yorkshire; an American calling himself Diophantes from Sirius (his own claim, not mine); Julian Butler, male hospital nurse; John Dunscombe, business director's son, and myself.

At 10.10 pm the attention of several pairs of eyes fastened on what Mr Butler described aptly as 'a burning bush on the ground' some six hundred yards SW of our vantage point on Cradle Hill. It was a fraction to the right of West Wiltshire golf clubhouse and a few hundred yards short of it, at a spot near a long and straggling hedgerow. Frankly, I first suspected it was rubble being burnt by farmer Geoffrey Gale (it is his land) yet we all commented how strange that this circular flame should erupt without warning at a late hour, no smoke or smouldering ashes noticed prior to this by our keen-eyed and conscientious observing group. Chris and Bob sped across the intervening land immediately, scaling a metal gate. Others followed at a more leisurely pace – perhaps because we were older.

The burning effect died on the ground and we were instantly aware of a large orange light, oval in shape, hanging stationary over the top of the lighted clubhouse. It was at low altitude, generally estimated at a maximum of a hundred feet. Glowing steadily with no hint of a surrounding aura or force field, it remained immobile for a full three minutes, according to my watch and supported by other time-checkers in the assembly. John said: 'It is far too big and brilliant for Mars, although it is practically the same colour.' Julian remarked pointedly: 'No, Mars is well away to the left and much higher, over Copheap.' There was no doubt left in our minds when the object, increasing its brilliance, started moving to the SE across Copheap and away over the shoulder of Battlesbury towards Starr Hill.

It was enormous as it moved slowly, almost sedately, throwing off a bright and fitful halo around the main body of the craft. We could now pick out a second similar-shaped object, much higher than the first and therefore smaller and a dull matt white in hue. It kept pace with the bright orange UFO, tailing it. But our eyes were hurriedly withdrawn from the double UFO treat. I had hastened back to the main body of watchers, anxious that one in particular (over 80 and quietly thrilled by the whole experience) should not miss the glittering spectacle. Most of the observers viewed the smaller spacecraft easily with the naked eye, without recourse to powerful binoculars that we carried. (Bob and Sybil had gone off with our three-inch telescope.)

Shortly after came the interruption. We heard cries coming from the field. Tearing towards us with ashen faces and trembling limbs, Chris and Bob Coates appeared. They vaulted the gate and were obviously agitated. Something had quite plainly upset and unnerved them both. They sat on the stony road surface, gulped draughts of hot coffee to recover from shock, and blurted out an astounding story of a near-encounter with an unknown entity! Little wonder their composure was shattered.

When they reached the border with Kidnappers' Hole, where the hedgerow terminates, they saw the burning bush peter out; and in its place was a tall figure dressed in a tight-fitting dark suit that had a peculiar silken sheen reflected in their torchlight. Bob was conscious only of 'something shadowy there', but Chris was more exact: a gold sash or bandolier (his terminology) was around its neck and shoulder, wound around the waist. No words were exchanged at the confrontation. Bob is six feet, one inch, in height. Chris thought the figure to be a good foot or more taller than his companion. Long, dark-gold hair fell to the shoulders of the stranger. It had bright eyes, colour not determinate in the lighting, and a rather feminine set of features in a not un-attractive face, Chris testified. The figure was motionless, one arm raised. Chris was unable to approach nearer than thirty yards from it. Overcome by fear of the unknown mixed oddly with an indescribable emotion, he and Robert decided to retreat hastily! Courage and nerves failing as the torchlight probed and splashed around the tall form in the darkness, they ran back to the hilltop to report in stumbling phrases what had happened. The two were questioned closely by the rest of the watchers while I made my way over the field, now bathed in moonlight unsullied by cloud wisps.

I had a flashlight and beamed a friendly message in Morse Code in front of me to allay and relieve any apprehension on the part of our unexpected guest. Then the futility and fool-ishness of this type of reasoning struck me. The visitor wear-ing the gold bandolier from his right shoulder feared nothing! There's no doubt at all about this. Behind him was the lighted clubhouse, with several dozen members within. On the hill were eleven persons, apart from Bob and Sybil out at Starr Hill. And, just in the dip towards Sack Hill and Battlesbury is the crowded School of Infantry, where there were thousands of troops not farther than a thousand yards from the Ufonaut. It was the first night of the full moon and everything stood out frosted clear, the copse silhouetted starkly against the starlit sky.

What bravery and calmness – and cool effrontery – the stranger displayed in face of all these worldly factors! He intended to be seen and came in a spirit of fearlessness because he hoped to make contact. Nonetheless, I admit that my knees were virtually knocking and my heart pounding more strongly than usual as I walked across the moon-bathed field; yet there is a scant hope of reaching understanding of the unknown in our very midst when fear is allowed to oust a spirit of genuine love and concern for all others in an immeasurable Universe, even the tiny speck that constitutes our own planet . . . I drew a personal blank on that lone journey. Nothing further was seen of the figure from then on.

Chris, Bob, Julian and John revisited the spot near Kidnappers' Hole where it had been standing on a small mound. They saw the hedgerow, clubhouse, two trees on the skyline; yet none could smell smoke or ash residue that an ordinary fire would have caused. So the burning of rubble seemed most unlikely – the farmer confirming this next day. No grass had been set ablaze, yet thirteen witnesses who saw two strange aerial objects failing to fit into conventional pockets of known aero-design, noiseless and sailing slowly through the atmosphere until spinning at faster speed upward, are far more convincing than the testimony of one poor soul on his own, often from then on the victim of merciless and almost unceasing verbal and written crucifixion!

On the night of 8-9 November 1978 thick fog spread in amorphous patches and clamped icy fingers over the sleeping Wiltshire countryside; it was almost hostile to those in search of cosmic truths, yet it proved to be memorable for the watchers at Cradle Hill. Engaged in a night-long sky-watch there despite the unfavourable weather conditions were veterans Chris and Pauline Trubridge, regular stalwarts on these excursions, plus two Australian lads from Sydney, Paul D. Davies and Ian W. Gale. Ian is a carpenter on an Australian rail network; Paul is a funeral director and embalmer who strongly believes in reincarnation; Chris is a joiner and music teacher (he plays the violin beautifully); while his wife Pauline is employed by a Warminster fuse-making concern. The two men from 'down-under' and I had chatted throughout the afternoon at my home in Portway and quickly discovered that our views and general opinions about our favourite subject, Ufology, tallied pretty well and amicably.

Because of a bout of temporary ill-health following a major operation for a nest of stomach ulcers, I was unable to join them at the hill on that eventful night, but all four contacted me next day to blurt out an extraordinary story which had the unqualified support and convictions of the whole quartet. In brief, the highlight of the vigil came during the early morning when they were huddled, bitterly cold and disconsolate, near the main white metalled gate fronting the military road up to and running alongside a copse. Sounding chillingly eerie in the still air, beating at the eardrums in the solitude, there came the distinct clanking of a metal chain from further down the hill approach, in the direction of Parsonage Farm. Paul flashed a powerful torch-beam full upon a couple of odd-looking intruders into the sky-watching venue.

Wearing what apeared to be heavily padded winter clothing and a cloth cap over his eyes in Sherlock Holmes' style, the man had a fairly long and straggly beard. But it was the huge dog, lumbering by the human's side in loping strides, which most caught the attention; it was somewhat shaggymaned, with a light-brown coat, and was clanking the metallic links of the chain connecting it to its companion.

The fangs were bared, but not fearsomely; nor was any growling or panting heard as it neared the amazed group. The four watchers behind the stabbing torch light boomed out a welcome to the man and beast: but only the rhythmic and repetitive clanking of the chain answered them from the darkness and swirling mist-banks eddying about the new-comers. When the pair drew near and stood silently in front of the sky-watchers, who were intent upon absorbing as much detail of the nocturnal pair as they could, mystified by their absolute stillness and silence as they stood on the shadowy road, something very dramatic happened which not only burnt an indelible memory on the minds of the four witnesses but served to amplify my contention that UFOS and their crews move and thrive in dimensions that are at variance with our own physical structural forms. It shows that they can and do come among us when they wish, and can appear and vanish at will, having mastered the advanced art of controlling atomic and molecular composition.

Just as we had seen UFOS plunge rapidly and noiselessly into sides and shoulders of Battlesbury and Cley Hill (as related elsewhere in this work), man and dog abruptly turned left and faced the high bank at the edge of the road. The bank is steep and topped by barbed-wire, in order to prevent cattle escaping from their field above the watchers' heads and into the road below. Then, before eight horrified eyes and four bewildered minds that seemed to be losing their grasp upon reason and reality, man and animal simply glided into the bank and vanished from sight! Paul told me, poker-faced and eyes still rimmed by lack of sleep: 'They made not a whisper of sound, apart from the clanking chain. They did not ascend the bank: just went into the side and out of our vision. It was incredible!' A startled Ian said: 'They were right beside us and nowhere near the barbed-wire when they went. They merely walked or glided right into the hillside and disappeared no more than a couple of yards from where we all stood astounded! It was decidedly uncanny.' Chris and

Pauline agreed that dog and master seemed to slide into the green grass of the lower face of the bank. 'It was as if a hole opened up and swallowed them from sight; but there was definitely no aperture visible,' said Chris; 'it should have been truly weird and frightening, especially with all those thick belts of fog around at the time; yet it was very noticeable that there was a lovely, haunting scent of roses around us from the moment we first spotted the bearded figure and the dog. What's more, the air in the vicinity became warm and comforting. I don't wish to sound soft and sentimental, but there was something really beautiful – one might almost put it as heavenly – about the whole unforgettable experience! I now feel quite certain that life-forms of this nature are often among us when we are sky-watching in this district – but *invisibly* so!'

═══

7

STARSHIPS
AT STARR HILL

A cub mistress and two cub scouts from Southend-on-Sea, Essex, and an observation group of twenty-one people assembled on Starr Hill in the farm barn area on Saturday, 22 April 1974. It was a clear and starlit night, ideal for viewing, the only mild discomfort from a fairly keen north-easterly breeze that made overcoats and flasks of hot coffee real blessings. A quiet night was nonetheless memorable for no fewer than nine lighted objects that were seen by every skywatcher in the space of forty-five minutes. The vigil lasted from 6.30 to 11.40 pm, yet the advent of nine glowing spheroids came in the tautly exciting period between 10.00 and 10.45 pm. At one particular striking juncture, there were three in the air at one and the same time, visible to all.

Every one of the lighted objects traced the same aerial path over the barn, flying in the direction of Stonehenge from Glastonbury. One aeroform was seen through binoculars to be carrying a light both fore and aft; green and red respectively. The objects were silvery-white, their altitude difficult to assess with complete accuracy, but the majority were flying well below flecks of cloud at some ten thousand feet. All satellites? Very hard to credit, especially when a trio of these glowing circles appeared in the sky simultaneously. Moreover, man-made satellites do not carry lights fore and aft! The aeroforms, noiseless and some speeding much faster than others which carved exaggerated loops in the heavens, were like pearly buttons floating on high. Again through binoculars, those nearest to observers on the ground had a broad rim of bright light, then were opaque within except for the centre, which flared into a living jewel of brilliance. A few pulsated in flight, both irregularly and intermittently.

Suffer little children to come to the fountain of truth! Having healthy curious minds and plenty of enthusiasm, backed by sensible doubts that they would see something savouring of the supernatural, ten young French students at school in Sutton, Surrey, expressed a wish to join our sky-watching team at Starr Hill, Warminster, on the night of Saturday, 22 June 1974. We welcomed them, for enlightenment is an essential criterion of UFO determination: the eye must see before the mind absorbs reality. Making the first overture through the post was Baron Arnaud de Gove, writing from 100 Mulgrave Road, Sutton. Together with the English tutor, Mrs Nina Bartholomew-Allan, the following boys and girls accompanied Arnaud to Warminster in a Land Rover and a car: Viscount Geradon de Vera de Margeliza, Marc Puechberty, Reny Yann, Beatrice Hatzfeld, Jean-Pierre Antin, Eric de Vaulchier, François and Pierre Gerard-Hirne and M. Dujardin.

They arrived in the town during the afternoon and, after tea, we took them to Cradle Hill to experience the atmosphere there, walking through the noted copse clearing near the golf course and discussing worlds of the unknown. An hour before sunset, we made our way by road to Starr Hill, where a number of sky observers were already assembled, including Neil Pike, Chris and Pauline Trubridge, George Woods from Lancashire, Bill and Jenny Yeadon from Westbury, Andrew Pritchard and several people from Glastonbury. Others, from various parts of the country, joined us later. For a while the sky was overcast, even on a warm summer night, but stars began to speckle the heavens at around 11 pm. Between that hour and 1 am on Sunday, we all saw six fast-moving flying lights, three travelling north to

72

south, the others west to east. Three were definitely ruled out as satellites, yet the remainder were mystifying; not so much to regular watchers, but to newcomer friends.

One UFO headed north in a distinct snake-like motion through the air. It was a pearly discoid, bright and exciting to watch. Another, which came straight in from the horizon, over the farm barn, suddenly soared upward and was in sight for about two minutes before abruptly vanishing. This was a dull-white and, under binoculars, appeared to be ringed by an aura or surrounding force-field of less light intensity. The undoubted fact that it flew upward before blacking-out proved convincingly that it was no man-launched satellite for these usually follow the curvature line of the earth. All six objects or lights were silent. Even at safe maximum flying altitude, at night, jet aircraft always leave a legacy of sound in their wake, so none of these was an earth-designed aeroform. A lady from Wells, Somerset, who conversed with our guests in French, told us she had seen a satellite regularly go over her house at 11.10 pm. But there was no satellite within fifteen minutes of that specific time on this night; not at Starr, anyway. We had a total of over sixty enthusiasts on the hill by midnight, including a radar operator and an electronics engineer, a music teacher and an ex-town councillor.

None was disappointed at what was seen, all agreeing that we had indeed caught more than a fleeting glimpse of the inexplicable chariots of the heavens. But let us study a few remarks made by our French student friends, who thoroughly enjoyed their first eventful sky-watch. M. Dujardin said: 'In August I am going to Greece and I will ask if they have UFOs – and I will tell them I saw some.' François Gerard-Hirne: 'I didn't believe there were such things as UFOs, but I cannot deny that I saw lights moving in the sky! I am off to Chile tomorrow and I will ask many questions there and tell them what I saw in Warminster.' His brother agreed: 'Before I refused to believe that UFOs existed; now, I accept that there are lights moving in the sky. Maybe one day I will accept that they are UFOS. I am now very interested in UFOS, I must admit.' Beatrice Hatzfeld: 'I understand all your stories and find them very interesting. I still can't understand how UFOS work and where they come from; but I was very happy to see the lights in the sky moving very fast. Thank you for a lovely day!' Jean-Pierre Antin: 'I wish I could have stayed several nights on the hills, and seen all the things you have seen. How can I thank you?' Reny Yann: 'After listening to you, I ask myself many questions. But whatever people say, I know I saw six moving lights. I am going to France and I shall make my own enquiries – and I hope to learn more about UFOS.' Baron Arnaud de Gove: 'I will mention all this to my friends and family. During my holidays in France, Corsica and Germany, I will try to find out about UFOS in those parts of the world and I'll let you know in October what I find out. Thank you very much for receiving us so nicely.' Marc Puechberty also promised he would speak to his friends and family of his experiences at Warminster that day. 'Maybe one day they will go to Cradle Hill or Starr Hill and see for themselves,' he hoped.

The young Viscount Geradon de Vera de Margeliza, who does not speak much English, wrote later: 'I enjoyed my day with you; and today I was able to understand what was on the tapes as our tutoress translated them for us. Thank you very much.' Eric de Vaulchier: 'Spending the afternoon and evening with you made me realize that there are many things I do not understand, but I was very interested.' All the students penned letters to thank me, afterwards, but we thank them in turn for at least bothering to come along and see for themselves to their own satisfaction.

On 1 June 1974 I lectured a friendly group of Aquarius Society members in a hall at Glastonbury. They liked the idea of a Starr Hill sky-watch, so came along the following Saturday 8 June. Their large coach managed to negotiate the hump-backed railway bridge near Starr safely, and I felt sorry on their behalf that the sky and weather were not clear. Yet a break appeared in clouds to the south-west at around 11.30

pm, and we were quietly thrilled to note that a rapid pulsation of blue light flickered for some three minutes behind each cloud 'curtain'. Then, at 12.40 am on Sunday, a silvery sphere came over the barn, west to east, seeming to slow at one juncture of its swift flight well below cloud level.

I left the scene at about one o'clock, as I was tired after a sky-watch the previous night, but another UFO-type aeroform was witnessed after my departure. Society secretary Miss Barbara Crump wrote: 'We were all so glad that we had a chance of seeing the two bright chariots before we had to depart; but even without that, everyone seemed to have thoroughly enjoyed the trip, as it gave opportunities of meeting others who were interested – and time to talk and ask questions. Perhaps you heard that some of us saw two more similar UFOs when we alighted from our coach, going in diverse directions? One executed a right-angled turn, which seems to prove that they really were unidentified flying objects and not satellites.' Is it not pleasing and reassuring to learn that more and more ordinary people, among them former cynics and severe critics, are actually proving with their own eyes and other senses that UFOs are actual and factual?

Sightings, sightings, sightings . . . Mrs Phyllis Kavenagh was retiring to bed at her home, Well Cottage, Crockerton, to the south of Warminster, at 11.20 pm when she noticed through her window, towards Lords Hill, a brightly pulsating light. It rose upward and dropped downward alternately, moving from right to left. It was like a star to begin with, then grew larger and brighter, turning reddish in colour. The witness watched the glowing discoid for some twenty minutes before it winked out. Quietly, she admits having watched this type of unusual sky phenomenon on other occasions in the past year or so.

Her visual experiences are shared by Mrs Robert Dufosee, widow of the former chairman of Warminster and Westbury Rural Council. She lives in a farmhouse in Deverill Valley, and she, too, has seen ruby-red and silvery circles of light in the night sky over Lords Hill, she told me. 'They are lovely to watch, for they sometimes remain in the area for many minutes at a time.'

David Robertson, of Bratton, is regional secretary of the National Farmers Union. He is also chairman of West Wiltshire Golf Club at Warminster. When playing a round on the greens of the well-known course in 1974, he and his opponent were staggered to see a long, cigar-shaped aeroform over Cradle Hill copse. It was as though the object, which glistened with a metallic sheen, hung suspended there in space. It was motionless and oddly majestic. David, an ex RAF man, thought to begin with it was simply a glider. His companion took it to be Concorde, for the aeroform was so massively proportioned. But because it bore no wings, did not move and made no noise, although only a few hundred yards from where they stood on the tee, both knew their first impressions were erroneous. 'I have never seen anything like this before, although I have heard some strange stories about these craft from various people,' Mr Robertson told me seriously.

Frank Brake lives and farms at Dilton Marsh, between Westbury and Warminster. His land stretches as far as Upton Scudamore, two miles from our town. Now, sheep love mustard (or charlock, as some countrymen call it), so he planned a big patch of it at Dilton Hollow, which some locals know as Dilton Bottom. Inspecting his crops one morning, he was amazed to find that a near-circular mouthful had been consumed overnight from the 4 ft high mustard crop, which bears yellow heads and is a familiar sight in Wiltshire pastures. He even took a commemorative coloured photograph of the peculiarity, showing it to me as evidence. The earth, visibly disturbed by something extraordinary from the air, was bared and measured some 21 ft by 18 ft. No marks were around the bald patch that would indicate a machine encroaching onto this field, but the affected area was nude and inexplicably robbed of mustard which had been growing there the previous evening. Frank Brake was also chairman of Warminster-Westbury RDC, and an unflappable person who is not easily fooled. Down-to-earth yeoman of our county, well-known in agricultural circles over a wide region, he told me in December: 'I have not seen a UFO, yet feel very strongly there must be something in what so many people allege they have seen hereabouts. After discovering this unusual damage to my mustard crop, I suspect there is much truth in this flying saucer business, after all!'

And at Westbury, a few miles from our town, Ian Collis observed for five minutes a UFO shaped like a dumb-bell and white in colour. It started its aerial manoeuvres at 8 pm he told me. While from nearby Trowbridge, at 1.22 pm on 15 September that year, Peter Mantell reported: 'A circular object moving towards the south at terrific speed. It was jet black in colour and I estimated it to be twenty feet across and about 150-200 feet above ground level, its rim lighter than its body. The object cast a shadow on the ground.'

Perhaps the most astonishing thing, a mystery almost as mind-shattering as UFOs themselves, is that the apathy over them among the Warminster public is appalling. I write this regretfully yet truthfully. In the pre-Christmas issue of the local newspaper our few stalwarts on local hills warmly invited everyone to come along and join us on a sky-watch at Starr Hill on Saturday night, 22 December 1973. Only one newcomer joined our ranks, a poor response from the critics who refuse to credit that so many witnesses are indeed speaking truth when they claim to have seen these modern wonders in the heavens.

The newcomer was Vivian Lane, who lives in East End Avenue, Warminster. He joined seven regular watchers at Starr, and was visually rewarded by a convincing show of gracefulness in motion. The UFO, a golden peardrop in the sky, suddenly erupted over Mortar Clump. It flew from left to right, west to east, its elegant duck-bobbing movement seen by all for a period of several minutes on its aerial route in the direction of Stonehenge.

On a clear November night, Peter Dearsley, of Wynford

Road, Frome, six miles from the most haunted UFO community in the world, was walking his dog and looking towards the west when he noticed, above the line of rooftops and about a thousand yards away, a bright light come into view. It was a good deal brighter than the surrounding stars, he observed. 'I looked at my watch and noted the time to be 10.07 pm,' he reported. 'After watching the strange light for two minutes or so, it began to move in my direction and slightly to the east, towards Cley Hill at Corsley. As it passed over the roof tops towards me, I had a clear view of it and could detect absolutely no sound whatever, thus ruling out the possibility of it being an aircraft. It was a constant, brilliant white light, triangular in shape, with what appeared to be the top glowing a vivid red. It was not travelling exceptionally fast, yet its speed was constant. It suddenly dropped like a stone and vanished behind the roof tops. The duration of my observation period was around five minutes and the object had in that time travelled about half the radius of the skyline,' he told me later.

Because of singularly odd events on local hills, such as persistent tugs at the back of my jacket, warm belts of air that caress freezing cheeks on wintry nights, lovely perfumes that waft full at the watchers in reassurance when some experience leaves them tense and occasionally fearful, added to the fact that I modestly claim to have enjoyed an average of two genuine UFO sightings a week over the past fourteen or so years of quiet research study, I am not convinced that the majority of what we see hail from other physical realms of the universe. A minority may come from other planets somewhere in this or another galaxy in the vast Milky Way; but most of what I have seen appears to belong to another dimension or level of existence in universal structure. It strikes me as an age-old force belonging to our environs and surrounds. A force, an intelligence, that is far older than Man and has inhabited the earth world far longer than he. As my colleague Brinsley Le Poer Trench postulates, it may in fact have 'seeded' us originally on this planet and therefore possess a strong binding link with and parental concern for us. In biblical parlance, it is the Ancient of Days, the Invisible Host, the Holy Ghost or the Army of Archangels still among us; just as the same good book tells us plainly that 'there were giants on the earth' in earliest times. Why should they not still be with us, unseen and unsung, mostly, yet ever-present and powerful?

An illuminated flying object which emblazoned the sky from Bristol down to the South coast amazed hundreds of people who witnessed it on 16 April 1978. There were so many reports flooding in to regional news media, television included, of the phenomenon (or phenomena) that to mention them all would require a special full-length work to be written on these incidents alone! Attestors from Bristol, Cardiff, Exeter, Plymouth and many other towns saw a bright aeroform silently gliding through the darkened sky in that early Sunday morning.

It first appeared over the Filton area of Bristol, passed over the Kingswood and Keynsham areas and moved southward. The strange sky trespasser was sighted over Warminster during its flight, then made its gleaming way down to Exeter and eventually to Plymouth. One snag arising when trying to assess its physical shape and colouring from so many reports stems from the fact that several witnesses described the shining monster differently and presented differing details of its direction. Over Bristol alone, according to Ian Mazyglod (newly installed editor of *Ufo-Info* and living in Bristol) we had reports of the object as being blue-red, moving north-east; green-yellow, heading south-west; red-yellow, going due west; and green-yellow again, fastening onto a south-eastern line.

From all these conflicting accounts of its aerial perambulations, the general impression gained is that the skyform was initially sighted at 1.05 am over north Bristol travelling in an easterly direction, probably more south-east. It was described as bright silver with tinges of orange or crimson.

79

Practically all witnesses mentioned orange-red particles trailing off. It seemed to Mr and Mrs Wilkes, Filton, to be 'cone-shaped or triangular, flying with a level trajectory,' David Teakel, of Speedwell, backed by a Mr and Mrs Bishop, saw the aeroform as 'flat at the front, tapered at the rear and flying at rooftop height.'

A few minutes later it was sighted over Warminster, although times of the two vary by some minutes. Witnessed from Cradle Hill by Brian Edmunds and Peter Smith (Cardiff) at 1.15 am, it was 'cone-shaped, point first, and belching huge red flames of fire and showering sparks.' Meanwhile, Mr and Mrs M. D. Hale (Bristol) gave a similar version, both present at the primary local watching venue; yet Mick Oram (Essex) and Marc Brinkerhoff (USA) recorded their unusual UFO sighting from another sky-watch point near Warminster as being at 1.07 am and consisting of 'a comet-like object varying in brightness and shooting out sparks as it travelled along noiselessly.'

Marc, who was staying with Mick in Portway at the time, shot a few seconds of movie-film of what they saw. Mick Oram and Marc Brinkerhoff told me that what they saw on 16 April was definitely heading south-east. The course of the air form appears to become confusing, now, for at 1.30 am Mr and Mrs Gloor espied the flying object from a coach returning to Bristol from London, soaring over the top of the coach as they approached Windsor. It was 'very large and white, trailing fire from behind it.' However, something that was 'egg-shaped, trailing flames,' was reported in Plymouth at 1 am by Mrs J. A. Bailey and her husband; this testimony supported by Mr R. Hancock of Plympton, as 'a metallic object, stationary'; and Miss Luke, North Hill, Plymouth, as 'silvery and tubular moving south-west.'

Eye-witness testimony also came in from Exeter, one in particular from Mr and Mrs Hoaken, who saw the weird intruder at 2 am emitting 'multi-coloured flames, moving in a south-west direction.' So the obvious question is: Was there more than one flying object involved in the sightings of The

Unknown that spread over several hours? It would appear to be the case, for it was reported from 12.30 – 1.30 am over Bristol, heading in varying directions; at the same time over Windsor as in Bristol and Plymouth; and the assumedly accurate timings at 7 pm of 'a colourless cigar-shaped object' over Plymouth (Saturday) and also at 4.40 am (Sunday) as 'a sphere illuminated like a neon light, travelling south-east' from Plymouth. All these variations and timings cannot possibly be associated with the aerial gyrations of a single UFO. Most reports followed the visual impressions of the one moving south-east from Bristol, which would naturally account for the Warminster and possibly Windsor sightings.

But Plymouth? Sightings recorded here and at Exeter might well have been of at least one other aeroform. Furthermore, we noted how the estimated height varied. It was reported to be flying at rooftop level in some cases; five hundred feet in certain others; and viewed in retrospect collectively, there is a strong case for reckoning on two or more aeroforms being active over that remarkably comprehensive April weekend. The 'official' explanation for the UFO(s)? It was judged to be a bolide (a large piece of rock which explodes after passing through the atmosphere). Certainly, this definition of the object does conform with such a theory in a few instances; but at rooftop level? And what about the changes in flight direction emphasised by many witnesses? Meteors do not persist in the sky for any appreciable length of time; yet the UFO was reported for several hours!

The nature of the silent intruder into our atmosphere is assuredly a mystery, yet it seems most likely that there was more than one isolated UFO lighting up the heavens early on that eventful Sunday morning in April 1978. As Abraham Lincoln once intimated: 'You can fool all of the people some of the time, and some of the people all of the time but you cannot fool all the people all of the time.' That goes for all things that remain inexplicable in the midst of lives we lead here on Earth. At popular sky-watching sites around Warminster, one tries assiduously to remain unbiased when acting as a fulcrum in the centre of a seesaw of contention about UFOS, their place of origin and their prime purpose in our native air space. Listening to the 'hard nuts and bolts' devotees who acclaim science as their God and too readily assume that Man on Earth is the supreme epitome of creation in the whole gamut of universal intelligence, versus the equally vociferous fraternity who claim that 'God is the author of all cosmic mysteries', the honest truth-seeker is often left wondering at the insipid, weakened state of progressive scientific evaluation and upholders of spiritual values alike. Why on Earth (or outside it?) don't the two main adversaries compromise to the extent of common agreement that no totally valid answers have yet been given to relevant questions UFO intelligence presents to meekly questing minds; and no particular group or investigative association has yet forwarded an absolute solution in unvarnished black-and-white terms to a worldwide enigma that has persisted for numerous centuries? The final answers, surely, must come to persons as individual seekers; to those humble enough to concede that *there is a middle way*; and a natural universal law which affirms that without challenges of The Unknown to our limited consciousness we should become nothing more than warped physical, mental and emotional cabbages; instead of destined to be a whole (holy) unit in an immeasurable Universe and an integral part of the wholeness (holiness) of the Creative Genius who set everybody and everything – all natural processes, divinely blessed – into motion.

Quiet debate is healthy; heated argument is useless! There are so many sides to UFO phenomena that can be sensibly discussed when reasoned compromise and complete understanding of others' viewpoints are heard and assessed in an untroubled atmosphere; whether it be at governmental level in the political arena or the Cradle Hill type level of truth-seeking experiences shared so willingly with thousands at Warminster.

8

═══

THE OMINOUS PURSUER

A personal adventure may not come amiss at this juncture. On the night of 16 December 1974, I went for an evening stroll along Westbury Road, Warminster. It was not an ideal backcloth for a sky-watch, with low and angry-looking black clouds heralding the approach of a rainstorm. During my walk I glanced towards the tree-clad slopes of the downs to my right and was struck by a leaping tongue of fire that lanced upward from the darkness of the sleeping boles. I stopped and watched, thinking at first someone was burning rubbish in that sector of the woods. But the flames broadened, while retaining the same height of some nine feet; so that finally it resembled a rectangle with a brightly-lit top to it, slightly wavering and jagged-edged. It then appeared to be solid in form, straight-rimmed apart from the rippling furls of flame capping it.

Slightly mystified, I walked onward, still convinced that it must have been an ordinary outdoors fire, well controlled by the burner, probably some industrious gardener. I knew instinctively that I ought to investigate it more closely, but other subjects engaged my mind as I took the railway bridge homeward across a large housing estate for council tenants. Yet two nights later, it happened again. Once more I was tempted to inspect the 'fire' at close range; and again I decided it was really no concern of mine, although the rectangular flame was thrust upward in the same or a similar clearing in the woods. However, when it happened a third time, as I went along the same highway a week later, my curiosity got the better of me at last. I ran up a short-cut to the downs and approached the site of the blaze from the coiling driveway of a private house. When I reached the vital spot, puffing and panting after a brisk sprint, I was shocked to find no flames there and not a trace of burning embers or ash on the dry ground! But I did hear movements in the undergrowth to my left. There were small, scuffing sounds which I took to be natural noises from disturbed birds or small mammals. But above these ordinary sounds, scratching and scraping of soil and late autumn leaves that had fallen, I was aware of a heavier tread that went away from me and up a steep incline. It was the noise, definitely, of the UFO-associated character we have nicknamed the Invisible Walker, for want of a more apt description: a sound that has been heard by sky-watchers on both Cradle and Starr Hills.

I followed the thudding footsteps, seeing nothing, even though I was flashing my torchlight ahead of me and from side to side. Then, as I neared the top of the hillside, breaking through the dense fringe of trees to clear space beyond, I saw it: a golden UFO, gently rocking to and fro, aloof yet not unfriendly, no more than a hundred yards from where I stood. As I watched, it left its invisible mooring, tilted its length upward and sideways like a drunken balloon – then sped away at tremendous speed, finally vanishing near Calloway Clump. It was no longer wayward, but circular, flat and oddly purposeful. I walked along the upper rim of the downs, but saw nothing further. Only the shrill hooting of a night-owl mocked my progress. Three times I saw the 'flames' altogether; and they never made another appearance along that perimeter of the highway. A hypothesis is defined as a basis for reasoning without reference to its truth. When a basis for truth has been established, a theory becomes possible. If events bear out the facts implied by the theory, it remains for general experience to confirm the theory as the truth, no matter how startling it may be. Belief is a proof of nothing! There was a belief in the past that the sun revolved round the earth, until the truth began to penetrate even the thickest skulls. Flat earth believers are with us to this day. Yet the Greek philosophers, without the optical proof of photographs from space, knew our planet was a sphere, because a man in a boat sees the masts of a ship coming up from over the horizon before the hull of the vessel is visible. The proof is there before your eyes – just as UFOs are when we least expect to see these remarkable jewels of the firmament! Millions of people throughout our little world cannot all be mistaken or guilty of faulty identification.

Regular sky-watch companions of mine at Starr Hill are Reginald Skrine and his daughter Susanne, from Trowbridge. They were with me one late summer night in 1975 when a trio of burning bushes erupted on the gaunt sides of Battlesbury Hill, one of the earliest Iron Age forts around Warminster. Again, the rectangular pattern was prominent, also the distinctive frilled upper edge. We saw an unwavering beacon or column of fire pushing as though out of the bowels of the earth itself, in triplicate, when we were driving towards Starr Hill, after we had crossed a railway bridge beyond Bishopstrow Farm. We stopped the car and took a closer view with the aid of binoculars. Just three great pillars of orange flame, tinged with red. We put it down to the Army messing around or the farmer burning off stubble, smoothing out rough grassland. This was a different season of the year from my Christmas-time experiences along the Westbury Road, so we naturally concluded it had to do with post-harvest burning. Yet I often think, in retrospect, it is a pity we had not fully investigated this incident, too, for there was something oddly similar to the Westbury Road trio of fiery mysteries. What would we have found or failed to find? Who knows? Minor enigmas within the main UFO one, nevertheless, for to the best of my knowledge – and after a number of inquiries – no one had burnt anything at that fringe portion of the prehistoric mound.

The pyramid of uniform-sized and rectangular lights add to the mysterious UFO jigsaw puzzle pieces of the Invisible Walker, the Tin-Throated Bird, the Slithering Serpents of Sound crackling underground at noted sighting venues, the bizarre things that have happened to fool-proof, electrically-controlled targets on military firing ranges in the locality, described fully in another chapter. Stranger than fiction? Real life often is!

Scattered pockets of pale and lack-lustre stars formed a filmy gauze in the black velvet of the night sky as the young couple drove onward through the Wiltshire countryside; past thick clumps of wild hedgerow and vast expanses of rolling downs beyond. Then newcomers to the town of Warminster, and totally unaware of its fast-growing reputation as a Mecca for unidentified flying objects, Dennis Tilt and his wife were shocked by an unexpected and frightening experience along a lonely road that October night. Checking carefully on their testimony later, as a journalist, I firmly believe they witnessed a landing and take-off of what the world calls 'flying saucers' at Chitterne, a village straddling part of Salisbury Plain and lying a few miles in the general eastward direction of Stonehenge, from Warminster.

They then lived in St Johns Road and both were convinced that 'it certainly had nothing to do with the military in the area, nor were they satellites or low-flying aircraft.' Yet the incident left a deep impression on their bewildered minds. The nightmare began as they drove home from Basingstoke and neared the sleeping village comfortably cosseted in the lap of the plain at around 11.28 pm. The only sound disturbing the hush of night was the whirring of the car tyres. All was still, peaceful, somnolently serene. But, on the far side of Chitterne both noticed 'three flame-coloured lights in triangular formation' on the ground, spearing up from the hedgerow and belonging to a farm on their right. They looked oddly out of place in that quiet spot at that late hour, and seemed 'unearthly, somehow'.

Dennis stopped the car, curiosity aroused, and stepped out into the road to enjoy a closer view of the brightly glowing trio of spheres sparking the darkness with rings of fire. They were glittering orbs that shaped a perfect pyramid of luminosity. A trifle nervous, sensing that all was not as it normally should be along the night-cowled highway, a creepy silence blanketing the air with the cutting-out of the car engine, Mrs Tilt advised her husband to climb back into the vehicle and continue their journey, 'without interfering in something that does not concern us'. Although her eyes probed the blackness as avidly as his own to focus on the hypnotic glare blazing from the triangle of lights in the field, he obeyed the sharp apprehension in her voice as she called out.

They drove along a bit farther. When they turned their heads to glance back curiously at the scene, rounding a slight curve in the winding ribbon of road, they noted that the strange lights had fused together, the three now united into a single smouldering mass of flames. The object, huge and circular, had lifted from the earth and was hanging as if suspended in the heavens. Both testified that it resembled, at this joined stage, 'a frying pan without a handle'. It was hovering some fifty feet up, like some giant golden bird, looming large and ominous, deadly silent. Prompted by his wife, Dennis put his foot hard down on the pedal, sharing the distinct unease of his trembling wife at the moment acceleration was achieved.

For, as the car surged into full power and tore along the night-capped countryside, the couple kept looking in the rear and further saw that the spaceship did not change distance between them one iota, although Dennis was proceeding at top speed, car shuddering its frenzied way round corners.

The wife maintained a constant watch on the UFO, which gave off a fitful yellow-red aura in flight; and Dennis peeped backward from time to time when he negotiated twists and bends in the country road. Both found it disconcerting and a trifle frightening that they were unable to shake off their aerial pursuer for several miles of an eventful homeward run. Unerringly, the sky giant kept aerially abreast of the car, a celestial chariot that skipped through the air like a graceful ballet dancer rather than a lumbering airship. It trailed the car and its huddled occupants all the way to town.

When they eventually reached St Johns Road, the brilliant apparition overhead had faded from the gaze. The couple breathed sighs of relief as they garaged the car next to their home. Yet there was one more upsetting shock to come. When Mrs Tilt later went up to the bedroom of their children, she glanced out of the window as she straightened the curtains. She was staggered to see the flying object again for a few pulsating seconds! It was a ball of shimmering glory in the heavens, slightly pulsing in an irregular rhythm as varied colours flickered through its diameter, before abruptly changing into an elongated egg-shape and blacking-out as though it had never existed. Because it is so preposterous in the light of earth concepts, one hesitates to decide what must actually have happened when the UFO ceased to shed brilliant beams down onto the speeding car in the final stretch to their home.

The scintillating spaceship must still have been present yet out of vision, according to its dramatic behaviour at the curtain-close of its vanishing act. So it is permissible to accept that these unearthly spacecraft are able to black-out whenever it suits their purpose, mystifying people and provoking sober thought along darkened byways in remote surroundings, giving the false impression they are no longer present. Even while the Tilts (he is a building contractor and former Army boxing champion) were excitedly discussing the

Cigar-shaped aeroform at Upton Scudemore near Warminster (*Chris Trubridge*)

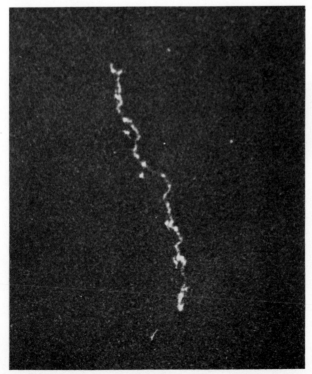

Night shot from Cradle Hill on 23 September 1978 (*James Rose*)

Another variety of UFO seen on the same night (*James Rose*)

Two unusual aeroforms spotted between 11.00 and 11.30 pm
on 14 October 1978 (*James Rose*)

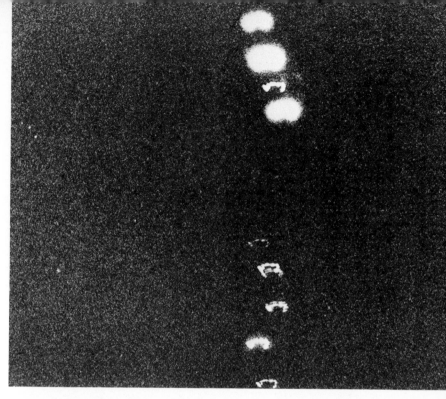

Amazing discoids seen gliding over Copheap on 21 October 1978
(*James Rose*)

Many witnesses saw these three flashing figures in October 1978
(*James Rose*)

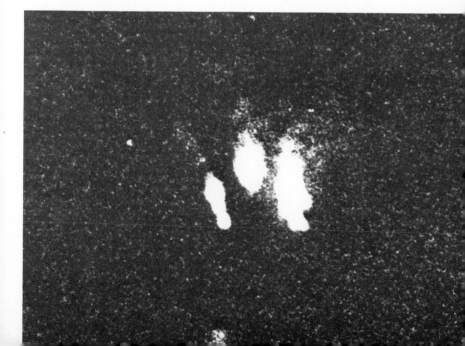

Nocturnal lighted spheroids moving at great speed (*James Rose*)

L-shaped bars of bright light which to some watchers on Cradle Hill
appeared to be triangular or pyramidic (*James Rose*)

An artist's impression of a UFO (*painting by Patrick Ford, 1978*)

In 1966 this UFO was photographed lurking over the rooftops (*Austin Reed*)

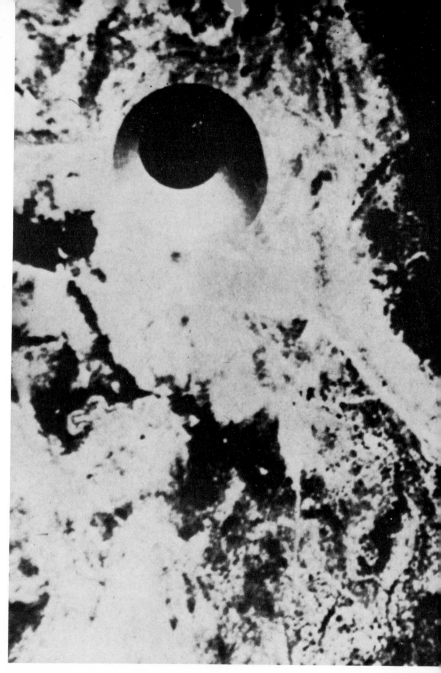

A pilot's view of a UFO (*Author's Collection*)

One of the Warminster-type UFOs seen over Dumfries, Scotland, in 1967
(*B. Sherman*)

vanishing act of the airborne giant, it was probably right overhead and invisibly dogging their car tracks on the last run in from Heytesbury to Warminster, unknown to Dennis Tilt and his wife.

So sure were the couple that they had indeed witnessed something highly untoward and absolutely incredible by earth standards, that Mr Tilt rose early next morning, which was a Sunday, and was at the farm making inquiries about the resting place of mysterious flying aeroforms that merged together before hanging hauntingly behind his vehicle the previous night. He drew a blank, however. The farmer was surprised but incapable of enlightening him with regard to what the 'one ball of light forming a triangle of spheres' was doing on his pastures long before dusk on its weird nocturnal spree. In spite of a close examination of the field and search of the surrounding area, the plastering contractor could not detect the precise spot where the spacecraft had taken off; nor were there tell-tale marks showing where the trio of lights had been shooting out their lambent light from soil level as the couple drove past. A baffled Dennis was only able to guess at the approximate region, with no indentations to guide him at soil level. His investigation, that of a sensible person bent on dealing intelligently with unknown and alien factors, proved to his own satisfaction that what he had seen was not even remotely concerned with military activity. He contacted the army and asked relevant questions, answers convincing him that troops on exercise were not the culprits responsible for the enormous airship.

Now for another true narrative, involving a case of two horse-loving girls and mistaken identity of a glider. Dawn Flanigan is a keen Warminster equestrian, living in Bath road, loving nothing better than a stiff breeze rippling her fair hair as she gallops and canters on horseback over the downs. She and her friend had several peculiar visual experiences in which their mounts also featured. Here is her description, given to me as a news reporter at her home before her parents, of patent UFO aeronautics:

While riding over the plain, in the general direction of Imber village, we saw to our left about five miles away a long object, silvery and metallic, that remained stationary over a wood. It lay towards the north, we reckoned. Frankly, we paid scant attention to it at first, as we are accustomed to seeing strange sights over Salisbury plain. The army and air force often carry out manoeuvres there, although we keep well out of their way when riding. The afternoon was warm and sunny, and the object glittered ever so brightly. The glint caught our eyes and we stopped to have a good look at it, then. It was a vividly flashing pencil of silver in the sun. About mid-afternoon next day, we saw a similar thing. But this time it was moving slowly over some treetops. Again we stopped and watched, the leaves a light green below the wingless fuselage. We expected it to twist and turn, or perform some aerobatics in the wind, as we now presumed it to be a glider.

But, as if our eyes were playing tricks on us, the thing simply vanished from view completely, only to reappear a few seconds later at the far end of the wood. Then it sank slowly down behind it and out of sight. The belt of trees seemed to swallow it up. By now we realised it was not a glider or anything of that sort, because it had an elongated shape like an old-fashioned zeppelin, was without wings or noise, and shone much too brilliantly in the sun. My girl friend and I changed direction and prodded our mounts towards the wood as fast as the rough terrain would allow. There were some questions we felt should be answered; and at that time we honestly thought the army were to blame, testing out some new contraption of war. However, on arrival there, we could find nothing to link with the funny air machine.

90

We also noticed that our horses were most reluctant to enter the tree belt, and quite impatient to leave the spot. They grew very restless and began snorting in fear and tossing their heads furiously, eyes rolling. Usually, they are calm and placid beasts. In that same month of May, we saw this sort of thing happen on several occasions. In fact, we investigated another three times. In each case we found nothing tangible on the ground – no traces at all of anything having been over or landing on that sector of ground. There is one particular occasion – the last – I shall never forget. That was when our steeds became so frightened that they reared up and bolted across the plain for home, in spite of our hard rein-tugging.

Needless to say, we had to give them their heads, and, sensing their alarm, we were not too anxious to remain near those copses too long either. Rather than be stranded out there in the wilds, we did not stop to argue! There are certain copses in the Warminster area that horses just will not enter. It is as though they possess a sixth sense, warning that danger possibly lurks in the undergrowth.

A Warminster foreman-fitter for a firm of agricultural engineers told me of his encounter with a 'cigar' type UFO one Sunday. George Radbourne and his wife Ruth live in Southleigh View. At 11.35 pm they had passed through crossroads at Chapmanslade and neared the straight stretch leading from Thoulston bends to Warminster, returning home after an evening out in Bath area. They blinked when a large airship hove in sight. It swept left to right across their car bonnet, from Warminster downs towards Cley Hill. It was white in colour but had a curious green sheen at its centre, the craft to the eye the size of a human adult hand at arm's length. Pretty large, as Mr Radbourne stressed when telling me: 'It jerked up and down slightly, in a rather erratic motion, then curved to low height in a graceful sweep, away to the right. The sharp-ended nose of the craft dipped and the whole structure blacked-out short of Cley Hill. It was silent in flight and seemed to bob about like a duck on pond water in its forward movement. Ruth, in the front passenger seat, corroborated his story. They too were comparative newcomers to our district and knew nothing of the town's amazing UFO history. Neither wanted publicity, to begin with, relenting when I stressed that names and addresses are important when evaluating testimony on a subject which, while important, is still largely futuristic and embryonic to the majority of people worldwide.

Feet-on-the-ground Mrs Ruth Radbourne told me she was not a bit frightened by the aerial vision. She said: 'It was somewhat like a weather balloon when we first spotted it. It maintained constant speed in crossing our path, and was doubtless under rigid control. The movement was not the same as that of a balloon and the object appeared solid at its nearest point. We assumed that it had landed, when the light went out. That was a mystifying part of the whole operation; how can something vanish in flight in mid air?'

Here is a story about cows that some readers might have deemed 'a lot of bull' at that period. This minor mystery hit national headlines in the late summer of 1967. A herd of several dozen cows disappeared from farmland at Chitterne one morning. In fact, they were absent from pastures and milking sheds for more than a day. An extensive search was carried out for the missing beasts by the farmer, his dairymen and other labourers, to no avail. The Army was indirectly blamed, one gathered, for broken fencing. Troops are often unfairly charged in this way. They denied responsibility, as no troops were in that vicinity at that time. Because of a baffling lack of tracks which should have been left by the fleeing creatures, the several hours of patient searching over a wide area found the cattle still 'missing parade', to borrow a military term. The hunt was called off by a worried farmer and his staff. Then, lo and behold, next morning the cows were all back in the field, closely herded together and lowing contentedly, as though nothing extraordinary had transpired to rip a page out of their calendar. A cursory examination showed they were unharmed.

Forty sheep vanished from a farm near Norridge wood (like Chitterne, also mentioned in my first book *The Flying Saucerers*), shortly after this. Despite an exhaustive search of the area for many miles, not a hide nor hair was found of the missing flock. It was assumed they had wandered afar or were lost in thick woodland between Norridge, Cley Hill and Longleat. They came back – or were brought back – to their grazing grounds a whole week later, wagging their tails behind them! Which raises a supposition: either we have abnormally adventurous sheep around Warminster, capable of flying into the night sky without wings, or they were deliberately abducted by aliens for necessary experimentation and organic inspection and safely returned undamaged and perfectly happy afterwards.

This leads us to observations made by David Holton, of Crockerton, near Warminster, registered homeopathic practitioner, surgical chiropodist and medical herbalist, who brought many early UFO mysteries to light in the neighbourhood. He believes the UFO manifestations are attributable to etheric energies. He says that the release of these energies from the soil, and more particularly from the soil of the tumuli, plague pits, graveyards and battlefields around Warminster, is the cause of strange occurrences which have mystified people all over the world.

He points out: 'Energy cannot be destroyed, only changed from one form to another, and this process going on in soil receiving the mortal remains of Man from the time he first appeared on this planet has resulted in an enormous build up of subtle energies which are released periodically for good or ill. This is very similar to a radioactive mineral slowly breaking down and giving off harmful radiation in the process. The energies released from the atoms of our mortal forms are modified by disease processes in life and by the

93

activities of such subtle etheric forces as mind at work through our entire being during life.' Mr Holton maintains that UFOS, or a large majority of them, are not material objects but etheric emanations appreciated by extrasensory perception. Therefore, they do not register their presence on any radar screens. Asked to account for noises reported by some witnesses from unusual objects in the sky, he says that one can hear at the ESP level as well as see.

Numerous witnesses of UFOS at Cradle Hill and Starr Hill can confirm that articles have vanished from cars and vans, even when vehicle doors have been locked – then miraculously turn up again when drivers and passengers arrive home. George Woods, a Merseyside area Ufologist who has braved inclement weather at Warminster many times, following long journeys south by train or coach, found on reaching home after his first visit that his tape recorder next morning had been tampered with, skilfully, by someone unknown. It had been expertly taken apart, components laid in reverse order on top of the mechanism. A silly prank? Barely possible. No one in the house had access to the instrument. I signed a copy of one of my books for a Birmingham Contact UK member. He placed it in his bookcase at home, locked the glass door and took the key to his bedroom when he retired that night. Next morning, he found the book missing from the shelves of the library. It reappeared several days later, in the exact position between two other books where it was originally placed! One UFO research group bought a recently-issued pop record while staying a few days in Warminster. Having secured it in a locked car at Starr Hill, the driver joined watchers at a gateway fronting a farm. He saw a shadowy figure silhouetted against the car window, *inside the vehicle*. Counting the people at the gate, he realised the figure belonged to no one in the observation party! So he ran back to the car and unlocked it. No one was then inside, but in the thirty or so seconds of indecision in which Neil Beverly enumerated the sky-scanners, someone had been in the vehicle. What's more, the new disc was grossly misshapen, wax-soft and depressed at the edges by obvious thumb impressions. It was warm and sticky, grooves spoilt and useless, ruined. A camera in the glove compartment by the dashboard had been dismantled, film turned around. Developed film had been taken from a container, replaced the wrong way round, and all manner of freakish things had happened to the boys' property in that amazing thirty seconds! Our investigating trio have encountered many mind-taunting puzzles like these, mystifying and challenging in the extreme, facing us at regular intervals, and have helped others search in vain for 'lost' property such as spectacles, knitting needles and wool, packets of cigarettes, small food hampers, coffee flasks, cameras, etc., which have invariably been 'found' eventually either during or after homeward runs of the losers.

We can only deduce from these perplexing conjuring acts that entities capable of manifesting in a duality of physical and spiritual forms, of solid and astral shapes, have been industriously at work to make disbelievers goggle at 'impossible' feats. Recent blatant examples are worthy of mention.

On 21 May 1977, three spherical and brilliant orbs of silvery light stunned a number of observers at Cley Hill, near Warminster, at 10.58 pm. After hovering in a static, tip-tilted fashion for several seconds over the lofty eminence, they abruptly changed direction and plunged pell-mell at blistering speed slap-bang into the solid side of the hill. And they failed to reappear!

And at Cradle Hill, three miles away and a favourite assembly point for sky-watch enthusiasts from all sectors of our native globe, the same extraordinary magic in motion was witnessed by over sixty people ten minutes earlier. Three white aeroforms, typically saucer-shaped or bell-like in appearance, burst suddenly through low-lying mist banks to the right of the earthworks at Battlesbury; hovered in ghostly fashion for a brief spell; then flew at lightning speed into the gaunt shoulders of the hill, seemingly to dissolve into night-shrouded slopes without emerging from the far side!

The patent fact that not a single watcher saw them leave the opposite point of the hill into a starlit canopy of sky above, left us all aghast. A repeat performance came minutes later at Cley Hill, mystifying beholders of the Great Spectacular Unknown that flies our skies. Nothing physical and tangible can emulate these bizarre occurrences. So: UFOS are decidedly other-dimensional entities in the universal framework.

Belonging to realms of spirit rather than flesh? Ultra-realists of the 'hard nuts and bolts' fraternity among investigators will protest: 'Perish the thought! It is unscientific and absurd.' But the true identity of UFOS, and their prime purposes in our hallowed atmosphere, will eventually become plain to each individual who is healthily curious about universal life-forms in a non-materialist and non-harmful way.

And perhaps they give us a timely reminder that matters of spiritual value came long before the first analytical reasoning and logic of science!

═══

9

INTRIGUING VARIETIES
OF UFO

Retired surveyor and chief public health inspector of War-minster-Westbury Rural District Council, Mr Frank Merrett was out shooting pheasants one Thursday afternoon on the Rye Hill Farm estate of Claudius R. Algar, former chairman of the local magistrates court and farming at Longbridge Deverill, a few miles south of Warminster. Shooting friends and beaters were with Mr Merrett at the time. A shadow uncurled on the ground, and the surveyor-sportsman imme-diately raised his gun into a firing position, expecting to see a covey of game birds winging overhead. But there was no whirring sound one associates with such mass flight. Above the farmhouse itself, casting its shadow in accord with the position of a wintry sun, was a long torpedo aeroform that shone grey-white and sparkled along its top where sunrays struck and bounced off its casing.

It was more rounded at one end than the other. What appeared to be slots or windows darkened the side facing the shooting companions. Frank told me bluntly, in front of a dozen people: 'I always thought you and other so-called witnesses of these phenomena mad, you know.' His eyes gleamed as he shook a wiser head. 'All I can now say is: if you are insane, I'm proud to join you! It was in sight for over three minutes altogether and made no whisper of noise. I ran along by the hedge to warn a gamekeeper friend in the shooting party, but when I reached him the object had gone.'

When he left his co-shooters, the silvery aeroform was moving over the house and slowly heading towards the downs to the north, he told me. My informant was quietly thrilled to have had his sighting of the unknown flying above us, I could tell. Other witnesses affirmed that the aerial torpedo continued its gentle coursing towards the downs near Cradle Hill, then vanished. One moment it was there; the next, non-existent. Changing of form is not restricted to night-time alien craft, either. Marion Bull, only daughter of *Warminster Journal* printer Vernon Bull, told me of a sighting when she was a fourteen-year-old schoolgirl on 28 July 1966. Between 3.30 and 4 pm, she was in Weymouth Street swimming baths when all saw 'a silver object, just above the trees near Ferris-mead. It was big and shiny, seemed to be flat to start with, then appeared to turn on its side. It then looked like a bowler hat with a slightly upturned brim. It kept going backwards and forwards, shape changing; now like a bone, slim in the middle.'

For fifteen years, Mr T. W. S. Dutton was vice-president of the UFO investigation centre in Sydney, Australia, in days when the talented late Dr M. Lindtner was president and the widely-read author Andrew Tomas was editor of their magazine, *The Australian Flying Saucer Review*. An engineer by profession, Mr Dutton was made an honorary life member in appreciation of his services, and he was a principal speaker on Ufology on radio and television on behalf of the investigation centre at Sydney. He wrote to me in May 1978 and admitted:

Like many other people, I was rather sceptical of the early sightings, but when I had a remarkable one myself it altered my views. I was flying to New Guinea (which I did frequently in my engineering work), and we landed at Townsville on the coast of Queensland at about 2.15 am. I went to the lounge, had coffee, then went outside to stretch my legs before the final hop to Port Moresby. It was a brilliantly moonlit night, no cloud, and as I turned in the direction of the distant mountain range I was amazed to see a bright reddish-orange oval object at about 35 degrees apparently over or near the mountains.

At that distance of several miles it looked to the eye about two feet wide and one foot high. I watched it for two to three minutes, then to my astonishment it swayed regularly from side to side for about a minute, stopped for a minute, then swayed again. It repeated this five times; after which it moved off to the right quite rapidly in a huge circle of some mile in diameter. It came back to exactly the same spot aerially, stayed stationary for about a minute, then began swaying again. In all, it repeated this manoeuvre five times before moving off to the right, and repeating the orbiting of the huge circle three times. Finally, when it came back it started to travel away from me at high speed, growing smaller and smaller to eventually disappear. I had drunk nothing stronger than coffee, I do not suffer from hallucinations and – as I said – I was sceptical of most alleged UFO sightings; but after that it was no use anyone trying to convice me I had been looking at a weather balloon, Venus, light or temperature inversions, etc! About two years later, again in North Queensland but at a place called Cairns, my wife and I watched another aerial object, same colour and shape but much farther away and therefore appearing much smaller. It moved from north to south in a slow, bobbing motion, through a distance of about 120 degrees horizontally; then it stopped for a few seconds only to drop suddenly and vertically through a distance of about 15 degrees downward. It stopped dead in its tracks at about 20 degrees above the earth.

As it dropped 'like a stone' at terrific speed it made a brilliant white flash in the sky. After staying still, motionless for a few seconds, it began to go away from us and soon disappeared. I could tell you of many sightings we investigated; some of them hoaxes, but others undoubtedly authentic and quite extraordinary. During my first sighting I felt a very definite telepathic contact; and I feel sure that there is a significance in the pendulum action repeated five times and the circular manoeuvre repeated three times. It has been suggested that the number 5 could indicate the planet in our solar system from which the object came; and that the 3 could indicate the time it took it to reach here.

I met George Adamski when he visited Australia and had a very active part in organising his lectures, etc. I also met Father Gill, who had that extraordinary experience at a mission in New Guinea, when he and over thirty natives exchanged signals with figures on top of a stationary UFO: and many other interesting people who are earnest seekers of cosmic and universal truths. We hope to have a final look at a few more places of interest in this fascinating country; and Warminster is at the head of the list!

He hoped I would keep up with investigations into the fascinating and unworldly events now occurring 'on our disturbed and troubled planet'; but another letter arriving by the same post in May 1978 came from the Ministry of Defence, Whitehall, and was in response to my cordial invitation for the Minister and other officials at Whitehall and in Parliament to visit Warminster any Saturday night in June to 'see for themselves' UFOs that often appear and flaunt their colourful facades between 9 pm and midnight; providing weather and sky conditions are reasonable. Miss G. J. Jamieson wrote: 'As you are aware, the Ministry of Defence interest in UFO reports is solely to see if they contain any defence implications. Investigations over a number of years by the authorities directly concerned with the air defences of the United Kingdom have so far produced no evidence that UFOs represent a threat to this country. Your invitation is appreciated, but we cannot extend our interest in UFOS by embarking on an independent scientific study of the phenomena, and we are unable to take up your offer.' Well, at least our government departments support my belief that UFOS represent no threat to the United Kingdom; but what a pity, while tacitly and discreetly admitting their existence and reality, the Ministry did not tell us *what* or *who* UFOS are! Maybe they just haven't a clue; so are leaving investigations in the capable hands of truth-seekers at ground level? For we are volunteers and there is no cost involved for any Ministry that is government controlled. What's more, we are free agents and not tied down by War Office rules and regulations . . .

When your witness list includes police, pilots, soldiers, lawyers, doctors, professors, men of science, astronomers, senior teaching staff, engineers, school children, law enforcement officers and parents by the million throughout the world, no responsible news reporter dares write-off *all* their testimony on flying saucers. Nor can such prominent world citizens as US president Jimmy Carter, Kontiki expedition leader Thor Heyerdahl, round-the-world sailor and pioneer airman the late Sir Francis Chichester, or four workmen of the Queen's uncle, Earl Mountbatten of Burma, be classified among the 'crank brigade' that cynics brand people of high personal integrity who bravely record what their eyes have seen.

Spots before the eyes? That is a laughable alternative to fact and reality! How can anyone of sound mind see such spots in blazing colour at night, which measure at least thirty feet across and perform amazing sky manoeuvres with unearthly agility, from 'floaters in the eyeball'? In spite of a barrage of cheap scepticism from narrow-minded people over the past three decades, the banner of universal truth will ever unfurl to support the millions of hearts and minds of sincere seekers who freely attest to these phenomena, earnest souls searching for the hidden mysteries of universal structure that currently elude us. Mention of UFO presences in many of the sacred works of history, embracing all known civilizations on planet Earth over past millennia, proves there is nothing new, novel or potentially hostile about these mysterious messengers from other worlds. Among witnesses at UFO-haunted Warminster was a tough Royal Marine commando whose car was stopped by a marauding circlet of fire that rose from a field and 'paced' his vehicle for some distance near Lords Hill, three miles from the town, early one morning. In a trembling state he reported the incident to Warminster police, who recorded it in their log book.

An equally tough ex-French Foreign Legion parachutist, veteran of three wars, Willy Gehlen of Salop, saw a landed spacecraft and a crew member between 6 ft and 8 ft tall at Upton Scudamore, not two miles from Warminster, at 3 am in mid-September 1976. All manner of weird things happened to his parked car in a layby before the encounter; and when he grasped the top of a metal gate to converse with the white-clad stranger, he felt a distinct electrical shock run through him when the visitor flashed an orange-glowing lamp from his waist section!

Many similar startling cases connected with UFOs have been reported locally, as my seven books on the continuing enigma illustrate, with all witness names and addresses given. Schools at Broad Haven, Steeple Ashton, Wincanton, and in Yorkshire have provided lively UFO-slanted stories that are absolutely true in recent months. So the UFO 'pace' is hotting-up in the UK alone, as well as in Spain, France, Germany and various parts of the USA.

Senior Warminster solicitor and ex-magistrates' clerk Mr Fred H. Knight, master pilot's wife Rachael Atwill, bank cashier Neil Pike and wife Sally (only daughter of former detective chief superintendant John Rossiter, Wiltshire Constabulary), Rev. and Mrs Graham Philips and family, ex-head postmaster Roger Rump, Police Sergeant John Bosley, Police Sergeant David Perry and daughter Janine, solicitors' clerk Gillian Ephgrave, ambulance driver Phil Champion, PCS Eric Pinnock and Graham Hutter and the latter's schoolboy son, Sunday newspaper science editor Bob Chapman, and many prominent townsfolk have all witnessed UFO magic in motion around Warminster; among hundreds of others including John and Maureen Rowston and Barry and Judy Gooding (who run the voluntary UFO-INFO information centre at Trowbridge, with other centres established at Fareham, Southampton, Weston-super-Mare and Warminster).

And thousands of visitors since the modern advent of UFO intelligence on Christmas Day, 1964, have also seen manifestations of unearthly nature in our skies after making special journeys to Warminster hilltops from Australia, New Zealand, USA, Switzerland, France, Italy and the Holy Land etc. As chief inspector Maurice Petty (senior town police officer) admits: 'They cannot all be mistaken – and they don't fear ridicule, which destroys truth!'

Bringing our story of elusive flying phenomena up to date, a variety of multi-coloured UFOS were reported by reputable witnesses in the West-country over the 9-10-11 June period of 1978, especially at Warminster, surely the Mecca for inexplicable aeroforms in the United Kingdom. Among them was the triple 'close encounter' during an hour-long nightmare drive along country roads by a youngish Welsh couple and their small daughter. Arthur Pusey and wife Muriel, plus little Joanna, saw three sky visions that bewildered them and left the daughter curled up in a rear seat, terrified.

They live at 34 Westbury Street, Llanelli, Dyfed, south Wales, where Arthur is a cycle engineer and mechanic. The shocked family reached the crossroads at Longbridge Deverill, on the outskirts of Warminster on the road to Shaftesbury, at 11.50 pm 'Through the windscreen, we noticed a huge ball of blinding light,' Mr Pusey told me. 'It had a white centre and a bright fluorescent green outer rim. It flew at no more than seven hundred feet in altitude, less than half a mile in front of us.' His wife agreed.

'It apeared before us to the left, flashing, and went out of sight to our right. It was lightning-fast, in and out of view in mere seconds. It was startling and so unexpected!' And the second strange aeroform in Blandform area was a reddish-bodied UFO that hung motionless before swooping downward before them. Again, no sound as it 'duck-bobbed' in levelling out and vanishing abruptly. Its height was hard to assess. Muriel Pusey said: 'It was uncanny as it hung there, just a reddish hazy ball of light in the star-speckled sky. We saw a dark-clad figure hunched over a lightless motor-bike, shortly after, on our offside. Obviously, his engine had stalled, maybe due to the crimson sphere above it. But we were scared it was a man from space, to begin with!' He was crash-helmeted and wore dark leggings.

Near their destination of Sturminster Marshall, where they were to spend a weekend with Arthur Pusey's uncle, the shaken trio saw what seemed to be a crystal-white satellite riding high above the car. Arthur stopped to view it through binoculars and was astounded when it halted overhead, hovering and growing much bigger and brighter. 'It was certainly no satellite or low-flying plane. It made not a whisper of noise,' he testified. Muried added: 'It was truly brilliant and huge! When it expanded, it resembled a firework opening out into an enormous mushroom of glaring white light in the darkness. It was so close that the widening gush of light was all around our car. We were terrified!'

Orange-red sky spectres were observed by dozens of watchers from Cradle Hill on Saturday night, 10 June 1978, the first gliding elegantly over the tip of early Iron Age earthwork Battlesbury at 11.30 pm. Among assembled witnesses were Dr John Cleary-Baker (former editor of BUFORA *Journal*) and veteran researcher Arnold West; bank cashier Neil Pike; County Council officer's son Richard Gardner; Chris Trubridge; ex RAF man Adrian Martin from Hove; Reg Skrine and daughter Susanne from Trowbridge with two friends; Arthur Gibbons and David Moses from Crewe. Mrs Dorothy

Fear's son and some companions saw a similar sky giant over Cley Hill later, and Chris, Richard and Adrian saw a flashing aerial monster from their car in early morning hours of Sunday 11 June.

As Chief Inspector Maurice Petty (Warminster Police) noted it on Monday: 'All very mysterious and puzzling; yet consistent with reports we have received here since the end of 1964.'

The effect of a visit by a UFO can have very odd repercussions. *The Daily Telegraph* recorded a strange experience on 25 September 1978: two men investigating the sightings of an unidentified flying object claimed they found coins on the ground which had been bent. When they emptied their pockets they said they discovered that a further six coins had also similar markings. Mr Trevor Aram, 20, and his brother Brian, 32, both tyre fitters of Stamford Street, Awsworth, Notts, reported seeing a large illuminated dome near their home. They are now seeking the help of scientists to solve the money mystery.

That was one of the rare nationally publicised UFOs that month; but hardly a night went by in Warminster area when circular orange flight forms were inactive that late 1978 month. Mr J. Griggs lives at the Widbrook Arabian Stud farm at nearby Bradford-on-Avon, for example. Contacting me on 22 September, he told me of his first experience of the inexplicable.

It was at 8.20 pm on 5 April this year. I was on my way up the drive to our house when I noticed a light in the sky. At first I thought it was a plane, but when I stopped to observe it was still there hanging in the sky. Then it seemed to get larger, giving me the impression that it was moving towards me. I watched it for about 30 seconds: the colour was orange and it flickered slightly.

Seeing it grew so large, yet no larger in that time, I ran

inside to get a telescope to view it closer, but when I returned it had gone. I watch the sky constantly now. I read up on the subject and have been trying to follow up on the theory of ley lines and UFO flight patterns. Taking the base for my leys as the eight white horses, standing stones and mainly old forts, I decided to watch from home, then if there is any pattern in their coming and going I could notice it. Nothing happened for months, probably because I was watching the wrong place. The next sighting was on 8 August, on Widbrook Hill at Bradford-on-Avon looking towards Trowbridge. At 10.20 pm three golden lights flashed on all at once. They appeared on an angle, leaning to the left as I faced them. I noted the order in which they went out. First, the top one extinguished completely, then the bottom one; then the middle one faded out to a small dot similar to the effect of switching-off a TV screen.

Mr Griggs watched this happen through binoculars over a period of almost two minutes. My new friend related how a golden UFO flared into radiance at 8.10 pm on 18 September while he was watching a plane pass over a hill. The light blazed in the heavens in Trowle direction, vanished twice and repeated its illuminating manoeuvres. These disproved an initial impression that it may have been a flare; for the aircraft had long disappeared from view! 'I started to run down the road past the farm, but gave up, thinking it could be miles away by now. As I returned down the hill, about five helicopters appeared and were circling the area where I had seen the UFO. I watched them for 20 minutes; one went overhead four times; then they departed,' he added, before telling me of various episodes concerning UFOs and members of his family, including his mother. I liked his farewell remark that: 'Maybe the UFOs are coming more frequently; or perhaps we are becoming more observant and they were there all the time!'

Even in an age of technological wonder where American and Russian pioneers are launching space probes far beyond our satellite the moon, it seems to be outside the comprehension of many people that intelligences from another planet in our galaxy may be doing precisely the same thing. We tend to believe blindly, because so-called experts assure us that the likelihood is laughably remote, that there is virtually no life on other planets in our solar system. Certainly not a type that corresponds to the carbon-based human structures we recognise and are familar with on earth.

Countless UFO sightings have been meticulously logged and duly recorded by earnest bodies of voluntary investigation. These laudable groups do not need financial inducements, nor the backing of government-goaded scientists, to compile a wordy report that means little in positive terminology to all the visually stunned witnesses of the unworldly with receptive minds. Many genuine sightings have never been mentioned and publicly revealed. Perfectly sane testifiers, not given to prevaricating or making loose-woven statements, have presented me with their valid UFO experiences in the Warminster area alone, insisting they be treated in strict confidences and not be published. These persons fear the inevitable backlash of widespread ridicule and the character-scourging whip of disfavour, which means proverbial exile from neighbours and workaday colleagues. People with closed minds and scant imagination would be more willing to equate UFOs with 'respectability' if the existence of such phenomena was firmly established without scientific dissent. Reports naturally hinge on the unquestionable integrity and reputation of witnesses and there have admittedly been cheap hoaxes occasionally perpetrated in a pointless manner in past years. They have even befogged the cause of truth at UFO-haunted Warminster, misleading a lot of honest folk at times; but surely there are so many samples of clear-cut testimony worldwide that some of them, at least, must be classed as accurate by any standards, concepts and yardsticks of possible measurement?

Maybe we are prone to assume too readily that all human life must necessarily conform to our own structure, with silicon or multi-celled animation out of the question. In spite of voluminous amounts of evidence in sundry accounts accruing throughout history, other-dimensional forms of intelligence that may exist in differing realms of time and space are scorned by some anthropologists, archaeological experts, historians, biologists, theologians, physicists and alleged students of natural development. All, according to them, are merely allusions to astral and sub-astral entities, to ghosts and spectral shades, to phantoms and poltergeists, that have in a misty past merged with stories of giants, fairylike creatures, elemental spirits, manifesting in doubtful historical documentation whose reliability is suspect. Yet perhaps it is a mistake to cast them aside as inconsequential nonsense: and not *all* scientists are unimaginative and disbelieving!

Comparisons between certain UFO incidents at Warminster and abroad show marked similarities in behaviour patterns. No one would believe, for example, the number of times serious observers at Cradle Hill and Starr Hill have seen these flying magicians change the shape and size of their craft; and how, on occasion, they have been positively identified as low-flying aircraft; only to alter into UFOs and dwarf their sound-waves later. Out-of-this world exaggeration? Even after witnessing this brain-bemusing performance above our heads on lonely sky vigils at night, we were inclined to shrug such visual experiences away as figments of restricted night vision. The human eye is not as keen or capable of shrewd distance assessment, for instance, at dusk as at dawn. Daylight makes estimates better balanced visually, than when the cloak of night has fallen over dimly outlined hilltops. Few scientists now rule out the possibility of visits by beings from another planet. That is an encouraging sign, reassuring to all truth-seekers.

Dr Paul Davies, a theoretical physicist at King's College,

London, says: 'Most people who dismiss UFOS as a load of rubbish do not realise how convincing the evidence is in some cases. If you have five hundred reports, all describing the same thing that cannot be explained, then we probably have something new to science.' And Dr J. Allen Hynek, director of the American Centre for UFO Studies, points out: 'We have a residue of truly puzzling reports indicating some sort of intelligence that may be quite close to Earth. The majority of these close encounters with alien beings last for several minutes and witnesses recall such specific details; so it is unlikely they are hallucinating.'

When railwayman Lambert S. Davies wrote to me from Porth, Rhondda, Wales, on 11 April 1978, I learnt of a 3 am experience. He was with a companion when they saw 'a beautiful cigar-shaped object coming over Cardiff very fast. It was blue at the central point and red outside. Although going at terrific speed, there was no noise!' That visual treat was way back in October 1970, but he had not reported the sightings earlier, fearing public ridicule, until he entrusted his authentic story to me over seven years later ...

An ex-Grenadier Guards sergeant and tank crewman in the last war, now foreman in charge of motor transport at 27 Command workshops of REME, Warminster, John Cotton saw a massive cylindrical object tearing through local skies between the Minster Church and Cradle Hill one Tuesday. It was late at night and he was returning along the Bath road by car. Patently shocked and surprised at the unexpected aerial phenomenon, not crediting UFOS until that solemn moment pregnant with meaning, he jumped from his car to make sure it was not the bright glare of a falling satellite or the 'hallucinated silver cloud' he first mistook it for; and satisfied himself that his keen eyes were not deceived. A further confirmatory witness?

John Bosley, Warminster police sergeant who was among surviving paratroopers dropped on Arnhem in the last war, noted a glowing red discoid that sped the same evening

109

towards Warminster downs in the north. He rang me to report his opinion that what passed overhead while he was out on night patrol was no conventional aircraft or satellite. It moved without a whisper. Yet both the ex-Guardsman and ex-paratrooper of an airborne unit bravely faced public ridicule by coming forward to relate respective visual experiences of the unknown. The police sergeant showed the same brand of moral courage displayed by several police car patrolmen who in 1967 saw 'flying crosses' at low altitude and chased them at 80-90 mph along country lanes in Devon. It is a pity that pranksters, or those who allege that UFOS are the machinations of moronic hoaxers, are so spineless in revealing what they profess to know – or are completely ignorant of – about the reality of the enigma, no matter *who* or *what* they might in fact comprise . . .

10

AN UNCANNY GIANT

A tough German who served with the French Foreign Legion parachute regiment during hostilities in Korea, Indochina and Algeria, battle-hardened Willy Gehlen has lived in Britain since 1968. His present address is the County High School, Bishops Castle, Salop. Still an active parachutist, he frequently jumps with the Royal Marines at Dunkeswell, near Exeter. It is fortunate he has nerves of steel and a placid temperament, thanks to wartime training, for he spent what normally would amount to a night of terror to the average person, in mid-September 1976. According to details he supplied me, it happened on a roadside between Westbury and Warminster near the village of Upton Scudamore, not two miles from our UFO-haunted town in Wiltshire. After a week of holidaying and parachuting, he left Exeter one afternoon to drive his roomy estate car to the Army Parachute Centre at Netheravon, near Salisbury.

I arrived there at 7.30 pm and asked the duty officer if I could put in a few jumps the following day. His reply was non-committal, for the parachute instructor was not available until next morning. As it was getting dark by then, I had some refreshment at Netheravon and set out to find a suitable camping site somewhere in the area to await the morning and a chance to do several sky-jumps from 4,000-5,000 feet.

In case you feel my adventure is difficult to believe, I must stress I am a pretty good observer in aircraft, and as a parachutist you must be endowed with stability, a no-nonsense attitude and good eyesight, plus excellent re-actions in any unexpected situation or emergency. This is a *must*; otherwise you would not survive for very long in this sort of sport! Now, as I am not a very rich man, I am quite happy to sleep anywhere there is sufficient room to curl up and be reasonably comfortable. I simply could not find a camping site of any description after leaving Netheravon, so I drove around for miles in the darkness.

Finally I became fed up and pulled into a lay-by near a village. I got my blankets out and decided to sleep in the car. But the rushing roar of traffic along the road kept me awake, so I resolved to find a somewhat quieter spot if possible. Therefore I drove on for a further several miles to a place called Westbury.

However, I just could not see a convenient lay-by, so I turned around and went back up a hill, stopping at the side of the highway near a gate giving entrance to a field. I did not expect anyone to come along during night hours wanting to get into that pastureland. It was now around the midnight hour, sky jet-black, so after laying my blankets out and locking all doors from inside, I left one window open for some fresh air, curled up and went to sleep. I don't know how long I slumbered, but I suddenly awoke feeling a bit chilly. To my amazement, the back hatch-door of the estate car was wide open – yet I knew I definitely secured it earlier, I banged the door hard from the inside, locked it and climbed out of the side door to the rear one to make certain it was absolutely secure and unable to be opened from outside. Back to sleep again; yet not for long!

And when I jerked awake this time, the flipping door was wide open once more. Normally I sleep very lightly and hear the slightest sound, but I had not heard anyone open that door on this occasion either. It was uncanny and unusual. Under ordinary circumstances it would be an impossibility! For no one could creep up on me and catch me unaware. Jungle warfare teaches you many lessons.

I looked at my watch. It was just coming up to 3 am. Feeling unsettled and a little uneasy, I decided against more sleep but to have a cup of hot coffee instead. I always carry a small gas stove with my camping gear, so I lit the stove in the car and put on a kettle of water. To do so I had to dismount from the vehicle and open the back hatch-door to provide enough space.

When I tumbled from the car I heard a peculiar humming noise, much like a swarm of bees in flight. Frankly, I did not pay much attention to it because I could also hear from somewhere in the distance a goods train rattling along a railway track. It was while I stood looking towards the train sound in the distance, more dimly conscious of the humming nearer at hand, that I was aware of someone standing not ten yards away, in the field on the far side of the gate and near an old barn. The sheer size of this person – between 6 ft 8 ins and 7 ft in height – made me wonder, but I was not frightened. Well, I thought it was a farmer, and me coming from the heart of the countryside in Shropshire, I knew that there was a bit of sheep-rustling going on. So I thought it logical that this giant farmer was merely tending his flock and guarding his animals.

I could not see his face, nor his clothes, as he was outlined against the back street lights of a village some mile and a half away. I remarked to him something to the effect that I meant to park by the gate only for this one night, as I could not find a camping site. If he wished to get a tractor into the field, I told him, I would be only too happy to move a few yards farther up the road from the entrance. There was no answer to my politely-worded offer. Instead, he shone a sort of square-shaped torch at me from his chest, I reckon; but the light it threw out was dark orange. I thought to myself: 'He could do with some new batteries in that torch.'

Yet, on the other hand, I had a funny feeling that he knew exactly what I was doing and the predicament I was in, having to sleep at the roadside. He did not say a word, and made no effort to come nearer. I then thought to myself: 'Well, please yourself, but I'm not coming over to you, because farmers can get really nasty when it comes to dealing with trespassers!' I put my hands onto the iron gate to have a closer look at the tall figure and reassure him of my harmless intention of staying there until morning, and I then received a tingling electric shock.

Thank goodness it was only slight, but I took my hands off that gate like a shot! My thoughts were warning me: 'He has wired his electric fencing equipment to the gate.' So, not to cause any trouble or argument, I turned away and began making my coffee – and took no more notice of the big shepherd standing so stockstill and silent in the darkened meadow.

After all, I was on a public road, just blocking a single gate entry for part of a night. So if he wanted me to move away he would have to come to *me*, now. But after a minute or so, I looked up and found he had gone. Yet I heard the odd humming noise again; and this time I saw what resembled a farmer's hen-house lifting off from ground level and vanishing into the distance. The queer thing is that the hen-house was some five feet up from the soil and gave out a pulsating pink glow. 'He is going to burn his hen-house and is towing it away to a safe place,' I thought; although I did not hear any tractor noise and the whole operation was carried out in perfect silence.

Everything in fact was deadly quiet. Eventually I saw the pinkly-shining hen-house rising up into the air at about 45 degrees, only to disappear quite suddenly. I concluded it must have gone up a hill road that was invisible to my sight in the darkness of the night, so I climbed back into

114

the car and went back to sleep, waking up to the blaze of 8.30 am sunshine. I gazed toward the field barn and into the distance, but there was no 45 degrees hill anywhere. I was myself on top of the hill. So where on earth did that hen-house go?

It was a quietly chilling and spooky adventure for a level-headed parachutist of a crack regiment, who was still not convinced that he actually made contact with a Ufonaut in the Warminster area in the ninth month of the year at 3 am. The tough warrior told me later that some friends of his in Salop, to whom he confided his nocturnal adventure near Warminster, midway between our town and Westbury on a windswept hill at Upton Scudamore, assured him he had witnessed a UFO landing and an occupant of a flying saucer.

He was advised to contact the Warminster 'expert' to verify this; and I am afraid Willy Gehlen lost a wager of six pints to them in their 'local' when I was compelled to back their collective contention that it *was* an unworldly visitation he experienced that unforgettable night!

Now for an experience which upset the nervous system for some while of Miss J. McCormick, of Roydon in Essex, who wrote to me on 15 May 1977 to tell me: 'I felt I had to contact someone who would believe what we saw. We were staying with my aunt and cousin at Cley Hill cottage, Whitbourne Springs, at Corsley just outside Warminster. We had gone to bed on Saturday night about 1 am and I was awakened at around 2 am. May I say at this point I do not know what woke me up. I went to the bedroom window and at first saw nothing but darkness. It was absolutely pitch-dark in the garden, although on looking left I could see clearly the outline of Cley Hill itself.

To the right of the window is the start of the grounds of Longleat. I noticed, after what must have been ten seconds, a light which was shining from the right – I assumed

– on to the garage which was facing right. I turned to look in the direction from which I thought the light was coming, and noticed a 'heavy mist' patch hanging on its own about thirty-five feet from the bedroom window. I looked again to ensure I was not seeing things that really were not there; and this time I was assured! The mist got brighter; not *really* bright, but bright in comparison with the darkness of the garden. It lowered slowly and formed into a bell-shaped outline. The shape was very prominent, although the light was only *within* the shape itself, and around the edges it appeared to be very 'fuzzy'. It then intensified into a brilliant white for a split-second, and was gone!

The effect it had on me was shock! (I am not a nervous person as a rule). My fiancé also saw this object about a hundred yards away, on a slope that is within Longleat grounds. In all, the time must have been no more than five minutes (accounting for the time elapsing between my sighting and that of my fiancé).

If it is any comfort to Miss McCormick and her young man, we saw some unusual sky incidents on the Saturday night from Upton Scudamore; and many flying streaks of light over Cradle Hill heading from north to south – and it is doubtful whether all could have been classified as satellites. There were some dozen or more in an hour's sky-watch; and on two occasions, two were zig-zagging simultaneously while passing one another in the sky! We honestly wrote-off half the quota of sky visions as satellites.

Mr Terence King accepted a lift at Frome in the early hours of Wednesday, while travelling from his home in Norwich to the west country; and had been journeying in the lorry on the A361 road for no more than twenty minutes when he saw 'a large object travelling a few feet above the ground.' Mr King said: 'At first I thought it was the moon. Then it came down nearly to ground level and its light clearly lit up the fields below it.' He described the tantalising object as 'huge and egg-shaped' and added: 'As it got nearer to the ground, something seemed to come out of the top, until it was triangular.'

'The driver though it was a sputnik,' continued the *Western Gazette* (Yeovil) news report. Both he and Mr King were 'white and shaken up'. The strange aeroform moved along in clear view for more than a quarter of an hour before vanishing. Frome police said later the same day that they had been informed of the UFO, but things had been seen at Warminster the previous night; and Frome is but a few miles from the notorious Wiltshire town. On the night before the hitch-hiker's experience in company with a lorry-driver, our team saw a UFO altering shape from oval to triangular and back.

Hospital chef Rodney Mullins and a friend witnessed an identical craft at 8.50 pm shortly before our sighting. Duality of form, therefore, applies to the capability of spaceships as well as occupants, one can deduce. The triangle is frequently referred to in UFO reports and bears out the authenticity of our photography in early times of the local saga. It could be representative of the Great Pyramid, of an important land area around Warminster, of the man-made colossus at Silbury in Wiltshire, or the supreme importance of the Holy Trinity at the present auspicious and revelationary period in the history of man.

Even in a breathless and windless atmosphere, boughs and branches in tree-darkened Cradle Hill copse, Warminster, often sigh and creak as if bewailing onset of old age. Rust-fronded or green-fingered in differing seasons, the undergrowth rustles and whispers weirdly amid other nagging noises that stab the minds of sky-watchers when UFOs put in glowing appearances. Things that are fairly commonplace become extraordinary and assume awesome, sometimes frightening, dimensions set against the background of the

117

bizarre. On occasion, reliable testifiers tell us in solemn truth, incidents occur which are far too unworldly to slot conveniently into the realms of science or fiction!

Kindly and helpful; mischievous or hostile? Here is a typical cameo reaching us in January 1977, when a visitor from the north of England told us he saw three definite UFOS in a brief sky-watch on their road journey to Warminster district. But what happened later, on an icy night that clamped down at their destination, intrigued me as an investigator of a Great and Spectacular Unknown. Their baby was asleep in the front seat, cosy and warm in her cot. The couple slept in the rear of the vehicle, after placing a spare blanket over a still-warm radiator to keep the water from freezing. It was when their arms and shoulders were shaken fiercely during darkness hours, by some unseen entity, that they awoke to find that the blanket on the bonnet had caught ablaze . . . If unchecked, the fire would doubtless have spread with calamitous results to baby and car had it not been for the invisible watcher or guardian who tugged at the adult sleepers so violently to warn them of their plight!

Why the frightening Mechanical or Tin-Throated warbler, weird noises and the clumping footsteps of the Invisible Walker, reported at various periods by scared sky-watching enthusiasts? These odd occurrences (such as items missing from cars at Cradle and Starr, subsequently 'turning up again' after homeward journeys) are not the actions of aggravating entities in our midst; not simply symptoms of higher intelligence intended to scare humans to death. Surely, they are meant to test our resolve and nerves; to determine whether we merit re-learning age-old truths that once, in the peaceful and equable environment of long, long ago, sang a lasting lullaby of love and perfect understanding in all hearts throughout the universe. In short: do we yet qualify for the inter-galactic brotherhood of man to which they are guiding us in present times? Man still cannot live by bread alone – and the milk of human kindness is not the monopoly or prerogative of man on Earth! We obviously have much to re-learn by re-thinking of triumphs in past ages, and lessons learned by abject failures, too.

Before going on to illustrative case histories which rivet attention to the truth of UFOs, from all sectors of our globe, we have proof galore that the science fiction of today is the scientific fact of tomorrow. Much news at present smacks of predictive literature of the past. Test-tube baby experiments were foreseen in the *Brave New World* of Aldous Huxley. In his *Journey to the Moon*, Jules Verne gave us exciting reading that was fairly accurate in descriptions of the landing zone, space craft and space suits. He also foresaw the advent of submarines in his *20,000 Leagues Under the Sea.* Remember? H. G. Wells wrote about tanks and their capabilities long before those massive armoured machines first terrified German soldiers during the first World War. Time travel is a popular element in fiction; the prospects of life complete with fourth dimensional potential is perhaps not far distant. Telepathy and extra-sensory perception may in future become a highly developed part of our lives on earth. People already claim ability to pick up the thought-waves or ether vibrations of others far away. There have been a number of striking examples of disasters being predicted by persons nowhere near the scene of tragedy. Pre-cognitive experiences and fateful dreams are recorded in history, but remain a primitive puzzle to science and leading thinkers, including psychologists and para-psychologists in our midst.

Mrs Pat Finch, of Salford Close, Redditch, Worcestershire, wrote to me about what she saw in July 1976 at Starr Hill:

> We broke our journey back from Cornwall to sky-watch at Warminister and arrived at about 1.30 to 2 am. We parked in front of the barn facing the gate. While everyone slept in those early-morning hours, I scanned the sky closely, through the windows, as I felt too nervous to get

out of the car. It was the clearest night one could ever wish to enjoy and I saw what I took to be a very large shooting star streak across the sky, expelling sparks or flames behind it.

After quite a while had elapsed, I suddenly noticed that one of the dozens of very fine stars (no bigger than a pin-head) seemed to be very slowly 'inching' its way across the sky. To make sure my eyes were not playing tricks, I picked out the nearest 'normal' sized star and watched to see whether the moving one would pass it. Sure enough, it did! After this happened, I lost it amongst all the other minute stars in the jewel-spangled heavens. It was a glorious night!

It did not travel in a straight line and it was definitely not a satellite! In spite of this real eye-opener, I soon began to feel intensely disappointed that I had seen nothing exciting at close-up range, this being my third visit to Warminster and the first clear night with which we were blessed. I was quietly musing on this when suddenly I saw a bright flash of light directly in front of me. I blinked to wake myself up properly. It was gone! Then came another dazzling flash, on and off again.

It stuck me at the time that it was trying to signal to me, in answer to my thinking how disappointed I was. As soon as I took notice of this double-flashing light and shouted to the others to wake up, the object started to rise up higher and then move along to the right of an arc, until the outlines of the barn hid it from sight completely. The light-form appeared perfectly round and about the size of a ½p piece.

As it moved, it pulsated on and off and changed colour to red, green and back again to a brilliant white. I also had the distinct impression that it was spinning. It definitely was not an aircraft, as the whole thing was light and accompanied by not a whispering trace of sound. Although I have seen nothing since, there has been a recent UFO 'flap' in the Redditch, Droitwich and Malvern areas, with literally dozens of people reporting sightings of a forty foot long silver object, stationary overhead.

Pat Finch's evidence reassured me that much was taking place of an unusual nature throughout Britain. Might there be more than a stray hint of 'things to come' in a letter from Francis Hurley, of Highfield Road, Bournemouth, who cites a Milan scientist who thinks: 'Flying saucers not only use light-energy as a means of transit, but by varying the frequency and thus the wavelength they effect variations of speed and visible changes of colour as seen by sky-watchers fortunate enough to witness a sighting of a genuine UFO in the sky.'

E

11

ANOMALIES IN PROPULSION

Gravitational anomalies and related subjects seem to be indisputably connected with UFO characteristics. A new type of dimensional system which scientific friends of mine are working on includes hyper-space and time shifts. Findings will hopefully produce a better explanation for many extraordinary things around us than physics has as yet been able to do. Obviously, studies of force fields, electro-magnetic or static in nature, have to be thoroughly tested and experimented with; for even though both are constituents of similar phenomena, they appear to be divided by certain aspects in relativity; and thus present an important problem to be overcome.

The gravitational anomaly in Santa Cruz which Bruce Cathie refers to in his second work *Harmonic 695* does appear to be a reality according to one friend, Mr H. H. C. Graepel of Kinsale, County Cork in Ireland. He told me: 'The department of the sheriff in Santa Cruz very kindly confirmed the existence of the "Mystery Spot" and even gave the name and address of the owner. You may recall that in this spot a small person will actually appear larger than when placed outside the spot. I recently sent the owner a lengthy questionnaire which he promised to attend to; and this should throw more light on many of the extraordinary properties which seem to exist there. I was also able to receive a similarly interesting booklet on an identical spot in Oregon, of which you may have a copy if interested. Yet another spot I was able to check out personally to my satisfaction exists in Sligo, Eire. I am, incidentally, a fully licensed radio amateur.'

Always willing and ready to help such sincere persons, I put Mr Graepel in touch with Peter and Ben Martins, of St Andrews, Bristol, who have discovered various triangles of land in and around Warminster that undoubtedly equate and align with ley-line statistics. Peter was once a direction-finder in the Baltic and other regions of the Western world in Royal Corps of Signals duties; and son Ben is a keen amateur scientific investigator. Father and son make a good partnership, exchanging and comparing ideas and projects. A generating source within Cley Hill, Warminster, intrigues them in an investigation into anti-gravity, very high electro-magnetic radiation, magnetic fields, possible space-time continuum, use of various radio frequency transmissions, oscilliscope readings, the unified field and advancement (if and where feasible) on the Philadelphia experiment.

There is an emission of energy from the two humps of the hill, and as they have discovered, a 180 degrees out-of-phase signal, denoting extremely high frequency in the magnetic field. Frankly, they are on a strictly scientific quest, yet tolerant of the factor that may intrude upon or concur with things of true spiritual essence in universal thought and structure. Doubtless, if their experiments amalgamate to *prove something*, no matter how ostensibly impossible and sensational, if practicable they will announce findings at a later date, when our world is ready for certain knowledge. There was a disruption of time dimension in UFO sighting areas. Simply put, some sort of disturbance is caused as a UFO punches through the time barrier into *our* time. The effect is 'something like a bow-wave from a ship as it ploughs through the sea. Possibly, it dredges up past time, when heathen practices took place here, devil worship and the like.

It is, however, a very real force, sometimes one of evil, to be reckoned with.'

Not of an exceptional scientific turn of mind, personally, I leave our formulae-seeking friends to their research studies. Yet one fact strikes me, in quietly ruminating moments away from our beloved hill. You are presumably sitting down to read this book and fondly imagining that you are quite stationary. Nothing is further from the truth! Consider: your house is in fact sinking by inches; your country is tilting slowly east to west and drifting generally westward; your floor is travelling over one thousand miles an hour with the surface of the world; and planet Earth is swishing around the sun in excess of a hundred thousand miles an hour. A giddy prospect, to be sure! So dare we assume something really astonishing?

That all flying saucers employ a simple method of propulsion when it suits their purpose? They simply stop moving! This conjures up the intriguing thought that on some occasions, when they are observed to be journeying at 1,100 mph, they are instead stopped virtually dead in their aerial tracks; and letting our globe spin past them. Again and conversely, when they seem to be static and immobile to us, they could be floating along at a speed of over 1,000 miles an hour in order to keep up with the revolutionary turning of our planet. If you feel that these staggering possibilities are in any way contradictory, consider for a moment the following: The countryside rushing past the stationary train; the ocean churning under a stationary ship; just so long as it is static relative to the surface of the globe we inhabit. In terms of a gravity wave, the UFO behaves exactly the same way as a linear induction motor, with the same results (See Diagram 1). Note firstly the direction of motion; how it stops at the end; and then consider a gravity wave and the UFO, where the wave carries the ship past an object by bringing the object past the ship. From here we can go on to study the infinite gravity waves intersecting; each point

124

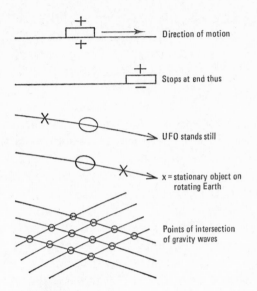

Diagram 1 (*Arthur Shuttlewood*)

of intersection gives a relative position of dimension. Thus we learn why the navigation grid used by UFOs is a constant overlapping of triangles – more properly, pyramids – presenting us with a clue to the entire business, as Stonehenge offers a calendar clock only and not a proper navigational compass in terms of a cone. It can be noted that the apex of every set of co-ordinates is based on logarithms and that units employed are constants. Why? Sheer practicability. We do not measure in micrometres but in large and easily handled units. Hence the use of activation lattices or areas bounded by specific interactions of curved gravity waves. They are the basic equivalents to longitude and latitude through three dimensions to give location in depth, linear distance and spacial co-ordinates: in more lay terms, a grand guidance system using a grid founded on electromagnetic wavelengths on planets – as ours does – and gravity waves in space. At last a comprehensive and scientific explanation for the grid can be offered. In no material manner does it contradict theories expounded in my last work on Ufology. Far from so doing, I would humbly suggest that these findings reinforce the arguments then used. They also supplement to a certain degree the ley-line findings of Jimmy Goddard and the orthotony ideas of Aime Michell; carried a necessary stage further because they are incomplete data in themselves.

Apart from the present-day controversy as to whether some prehistoric survivors are still submerged in the placid depths of Loch Ness, it has long been thought that the majority died out because their food supplies failed. The latest theory is that they may have been killed off by something harmful in a reversal of the North and South poles. It was fairly recently discovered that at least 171 times in the past seventy-six million years, the magnetic field of our earth has suddenly faded, then returned to normal with poles reversed. Experts say the changed magnetism in pole reversal could somehow have been responsible for the disappearing dinosaurs. In the past ten million years the poles have reversed about every 220,000 years. The last time was 700,000 years ago. Which means that another switch is overdue. So man could soon find out for himself what actually happened when the dinosaurs died. It is a rather sobering realization. Yet careful exhumation of ancient tales shows that ufos or flying saucers – which undoubtedly constitute another enigma – are as old as history. By way of starting a curve, it can be noted that they seem to come upon us unawares, like strange shock-wave troops or perplexing intruders in our hallowed atmosphere.

In 1887 there were numbers of sightings of these mystery airships in America. In 1907 two amateur astronomers saw what looked like airships crossing the face of the moon. In 1911 spaceships equipped with searchlights were seen over New York. London, city of some ten million people, was treated to a visual beauty at 9 o'clock one day in September of 1946. Eighteen months before, most of these good folk would have been casting glances skyward in hopes of seeing a vapour trail of a v2 that had passed over to explode else-where. But in 1946 apparently only one Essex man, his wife and niece, saw it for there was no mention of this aerial colossus, either in evening newspapers or on radio news. Therefore, although it appeared to their startled vision to be a huge aeroform, larger than four *Queen Marys* shining like snow in the reflected light of the morning sun, and so obviously rotating on its horizontal axis, they just could not credit the evidence of their own eyes. Once a ship's chief engineer in foreign waters, Robert James Bond, of Grays, was left to wonder . . . as were the two head-shaking women-folk. Only something immaterial, such as a searchlight beam, could cross the sky at 7,000 miles an hour at 15,000 ft and make no sound. Yet this was no mere light beam. Its passage deflected an upper cloud layer, which at highest ceiling did not exceed 22,000 feet, and pushed down a lower layer that was no more than 4,000 to 5,000 feet. This suggested that it was a solid body forcing the air upward. Yet no follow-up of a vapour trail was visible, as should have persisted for some while afterward. The lack of sound denied a solid mass. About forty-five million horsepower would be necessary to drive a vessel that size at such speed. The noise should thus have been severe and appalling! At that time the sound barrier had not been broken, although two days later a young pilot lost his life trying.

It seemed to be either a delusion or an unknown whorl in space, possibly the weird resultant of the Bikini Bomb of July. A fortnight later, twelve farmers in Arkansas reported as a group that they had witnessed a spaceship cross the heavens. The USAAF said they probably saw a B49 bomber which was on trials in the area. Later that autumn RAF fighters chased aircraft believed to be smuggling drugs over the South coast of England. They escaped from some of the fastest planes in the world! Even later, amateur astron-omers reported a radioactive cloud, thought to be from the Bikini detonation, visible for weeks at high level. It was last discerned at 40,000 ft. There were other news reports, all of odd incidents that defied logical explanation, which – taken together – indicated that a mass of particles having special properties was blown into space at the Bikini nuclear test explosion.

With touches of imagination, one can visualise that it remained intact for some period of time as a highly magnetic structure hit the English Channel in mid September and caused a massive tidal wave. Dutifully, the weather bureau said it was due to a temperature inversion. The mass then ricocheted, orbited the earth again, passed over London at low level, demolished a church in Australia two nights later (cause stated to be unknown); then, after the farmers saw it, the mass began shedding fragments and the first flying saucers were witnessed in the spring of 1947. Much remained together, seen by White Sands surveyors some while later. Bikini had fifty-one devices exploded on it in all, but no similar masses were evident because the air around had already been polluted to such an extent that it could not happen twice.

Einstein postulated mathematically that space is curved; and this was verified by optical means. He also predicted that space and time are linked in a continuous process. It has now been established that the spiral space-time continuum is composed of almost identical helium alpha particles, normalized during their passage from the sun, where they are being continually formed by the fusion of hydrogen atoms.

The possibility of nuclear fission, followed by actual use, led to the probability of nuclear fusion in the sun as its source of energy. Nuclear fission (based on the formula $E + Mc\,2$) separates the nucleus, a helium alpha particle, from its electrons. The rays coming off radium are called alpha, beta and gamma. The alpha rays are actually particles which are the nuclei of helium. Helium is an inert gas, but alpha particles are assuredly not inert; they are nuclei without any circulating electrons or sub-atoms, and they remain so until they attract enough electrons to become helium. These nuclei are the heaviest things, for their size, in all creation: the heaviest rock is lighter than a feather by comparison! They are also very powerful magnets and are drawn together to form swarms like tiny bees. Because of their weight, they are blasted further from the sun into outer space than any other atom. The swarm then forms and tries to grab as many free electrons as it can in the Van Allen belt before its weight forces it to gravitate to Earth. However, there are not enough electrons to go round and, apart from this, only the outside nuclei in each swarm ever get enough to cover one side before they are dragged into the centre by the magnetic force of the other nuclei. Where a mass of alpha particles is grouped without electrons to separate them, the entire mass becomes exceedingly unstable. This is because unsatisfied magnetic poles form where groupings of helium alpha nuclei are asymmetrical. The outside may become unsatisfied for a brief period, but all the while magnetic action is taking place within the mass and a fresh grouping begins. One grouping may resemble an airship and another a simple crystal. The skin or meniscus on the exterior covers a seething whirl of highly magnetized masses of unsatisfied alpha particles, and this explains the unpredictable behaviour of some UFOs seen.

We see these swarms when they have gravitated back to Earth and have by this time become larger; as the nuclei become separated by electrons the mass becomes lighter. A swarm in this state circles round Earth in the manner of a satellite, still trying to collect electrons to satisfy the nuclei. UFOs come down from invisibility, become temporarily normalized, then, repulsed by terrestrial magnetism, are forced to rise again.

Earnest UFO students will recall that the Condon Report scientist laughed off such objects and their explanations. They examined thousands of alleged sightings and all were derided except two. There was a tiny ball that acted as if under intelligent control or as though possessing its own intelligence. Condon grew curious about this little sphere having physical structure. Not one of the scientists on the panel had ever seen anything inexplicable in the heavens, a factor which really rendered their report invalid, but the tiny ball which only Service radar could 'see' could not be mocked. However, as it ocurred on an airfield in England, it could not frighten the big American public as Orson Welles' radio feature *War of the Worlds* did. As for intelligent control, all the 'tall stories' of Ufology lend colour to this. Yet, as at Warminster in England, there are glaring exceptions to the scientifically acceptable and explanatory logic of the alpha, beta, gamma proposition. The pilots and crews of these bizarre spacecraft have actually been seen. Which at least invites the suggestion that there is something humanoid, tangible and physical about the occupants.

Yet, apart from ghosts, phantoms, spectres, poltergeists and sundry imponderables that go bump-in-the-night, there are levels of existence and scales and dimensional aspects of life that still elude the horizons of our knowledge. No one knows all the answers to the perplexities and conundrums that current scientific concepts cannot measure. And UFOS come into this frustrating category.

All truth-seekers, especially in the specialist field of serious Ufology, ferret out information that consists of verifiable facts and loose-woven rumours that prove difficult to pin down positively. When assembled into the jigsaw puzzle that UFOS represent, however even such pieces of evidence that are deemed suspect by hardened investigators ('contact' allegations by Adamski, Villas Boas, Barney and Betty Hill and others) begin to cohese into a pattern of sorts.

They also make out a pretty good case in favour of contact having been made with at least one government. In spring 1954, base personnel at Edwards AFB in the Mojave Desert suddenly found that a security barrier had been established. No one was allowed to enter or leave the base for 48 hours. At the same time, President Eisenhower announced his departure for a golfing holiday in Palm Springs; although his valet let it slip that the president had not taken his golf clubs. Eisenhower *did* arrive at Palm Springs, I learnt from a reputable source, but whilst there he was spirited away to Edwards AFB by helicopter in the same 48-hour period.

During those eventful two days base personnel, in spite of a strict security blanket, late reported seeing five different types of 'flying saucer' landing and taking off from the base. They demonstrated ability to be visible or invisible; also to appear on and disappear from radar screens at will. Several small (about four feet tall) humanoid creatures were seen in discussion with assembled 'top brass'; and these senior dignitaries were witnessed entering the craft, taking rides in them. When the UFOs were soon after known to have vacated the base, the tight security ban was lifted.

Thus ended the 'incident'; yet the complete story does not terminate at that juncture. Within a short space of time, press censorship clamped down. From 1947, news organs had been crammed with tales of UFO sightings; but in 1954 this ground to a halt. All such stories abruptly vanished from the United States press. Censorship was not confined to the USA: it also involved England, Europe, Africa, Asia, China and the Soviet Union. Why should almost every country worldwide suddenly drop UFO stories like hot bricks? Our Earth was in the grip of the Cold War, mortar 'still wet' in the infamous Berlin Wall, both sides glaring at the other over it. How then could global 'co-operation' and brotherly love be achieved?

Is it possible that every major government received a similar visitation to the one at Edwards AFB and, following this, applied their own version of 'D' notices served on the press? Noteworthy, it was around this time the infamous and frightening MIB (men in black) appeared on the scene. Since when, we have continually been assured that UFOs just do *not* exist! The US government actually spent 300,000 dollars on the Condon project simply to state this in emphatic terms: yet simultaneously the same government maintains No. 4602 air intelligence squadron for the specific purpose of investigating the very reports the project denies! Consider the costs of maintaining a fully-equipped air force squadron against those incurred by the Condon project? Clearly, one can see where priorities really lie ...

The foregoing provides a fairly substantial, albeit circumstantial, case for the biggest and widest-reaching contact event of all time! It is conceded that hard proof of this incident is not open to public scrutiny, safely housed in the Priority One, Top Secret files at Wright Patterson AFB; as Senator Barry Goldwater discovered to his disgust and chagrin when, 1973, he was refused permission to scrutinize them! Goldwater is an ex-Air Force Brigadier and ex-Presidential nominee; yet even he was not allowed to see the file pertaining to Edwards AFB of Spring 1954.

There *was* a 48-hour security barrier erected at Edwards AFB at this period; there *were* reported UFOs in that area at that time; and there *did* ensue a worldwide press censorship of UFO stories. Facts about this incident may be termed hearsay; yet I have been assured that every fact can be confirmed should any truth-seeker go to the trouble of personal investigation. Unfortunately, even case-hardened reporters run into brick walls when it comes to unearthing facts about UFO sighting reports by Service personnel! So readers must judge for themselves. Doubtless sceptics will consider the case 'too fantastic to be true'! For my part (and that of faithful friends not given to telling lies or fabricating unimaginable stories along the UFO trail) when next I am asked: 'Why do not UFO intelligences contact any governments?' I must in all conscience reply: 'They *have* done so.' (APEN friends will agree with this bold assertion!)

Readers will have noted the stress I have placed on the possibility that our UFO intelligence is other-dimensional, rather than extraterrestrial. And the rather remote suggestion that they may hail from our future? In our accepted terminology, they are 'out of this world' yet part and parcel of it. This is not a contradictory assessment, however. It will be recalled how I emphasized the capacity for UFOs to penetrate solid matter: enter into the sides of hills and not come out: as if swallowed up by invisible rock ledges and mountainous peaks. Incredible, surely, unless these eccentric machines are able to transmute from density into a different atomic and molecular status? If the reader will bear with me a trifle further on this ridiculous and seemingly ludicrous theory, let me give a few more examples of how the idea first took a place of precedence in my research on the whole UFO enigma. Shall we take a deeper than cursory glance at a series of sightings that bring the UFO saga into fairly recent prominence, at the latter part of 1978? A newspaper covering the Hounslow, Isleworth, Brentford and Chiswick area of Middlesex told us in its issue of Friday, 1 September 1978:

UFOs have been in the skies over Chiswick, a number of people claimed this week. People living near Concorde's West London flight-path claim UFOs from another planet have come to keep an eye on the supersonic jet, one of the country's major technological advances. Eye-witnesses describe the UFO as a huge ball of red light, changing shape; exactly the experiences described by scientists and astronomers who study the subject.

First to see the mysterious flying object was Mrs Deanna Godden (66), of Edensor Gardens, Chiswick, whose flat faces the Heathrow Airport flight-path. 'I have seen something I'd never thought I would see in a million years,' she said. 'A fortnight ago, I saw a huge reddish ball in the sky the same time that Concorde was due to fly over. As Concorde came over, I was terrified there was going to be an almighty crash; but the plane passed right through it! My family and I saw the same thing on four consecutive nights at the same time when Concorde flies over, just before 9 pm. I know what I saw, and it is very strange.'

Porter Mr Jim Hale (50), of Bedford Park, was returning from a holiday in the West Country when a glow was spotted in the sky at the same time as Mrs Godden's sighting, over the M4 motorway. Mr Hale stopped the car and saw the object change shape five times rapidly, through a cigar, kite, disc, and square string-shaped form. 'I refuse to believe it was a star, because it changed shape so many times,' he said. 'Mrs Godden and I saw the same thing on the same night. I am convinced it was no optical illusion.' 'What is even more spooky is that I have since learnt that a glowing ball accompanied Concorde for part of her maiden flight,' Mrs Godden added. Those are the details in a boldly headlined *Chronicle* news story. Maybe the most relevant observation is the remark that '*the plane passed right through it!*'

This plainly contradicts any supposition that the UFO possessed any purely physical and therefore tangible properties; and it could justifiably be argued that it was other-dimensional and not necessarily from another planet somewhere in our vast universe . . .

12

SILENT SAUCERS
OVER STONEHENGE

I was especially pleased when one visitor to Warminster saw
something unusual. It could not have happened to a nicer
young man. Ian Girvan, whose father, the late Waveney
Girvan, was a pioneer in flying saucer investigation, author
and one-time editor of *Flying Saucer Review*, came up to
Cradle Hill one summer night, accompanied by his farmer
brother-in-law. Ian badly wanted to vindicate his father's
opinions and works concerning UFOs, so he was particularly
anxious to enjoy his very first UFO sighting that night. Through
a pair of powerful binoculars, which they quickly aligned
on to the first pulsating spaceship that came over Warminster, both saw no fewer than six other UFOs flitting in
rear of the main one.

'The large one we could see clearly without the aid of
binoculars,' said a highly delighted Ian, like myself a journalist and hard-headed realist. 'But it was only when we used
the binoculars that we could see the others far beyond, very
distant but discernible through the lens.' With the naked eye,
these half-dozen would have been invisible and therefore
uncounted. Bob, Sybil and I may have inadvertently been
missing scores of more distant UFO brethren, by not always
employing binoculars in sky-watches.

'Are UFOs still appearing over Warminster?' In answer to
hosts of inquiring letters I have received since publication of
my first book, *The Warminster Mystery*, the short and honest

reply is: 'Yes.' Here is one news story in the *County Press* carrying the front page headline 'Blinding Light Swoops on Car,' for instance:

The Warminster Thing has reappeared. Telecommunications engineer Mr Trevor Marsh, of 10 The Close, Warminster, used to have an open mind about the Thing, but he says after it hovered over his car on the main Warminster to Bath road near Warminster on Monday night, last week, he has no doubts left. Trevor was driving home between 10.45 and 11 pm after visiting his girl friend when he saw a flashing light over Norridge Wood. He thought it was a helicopter's navigation light. But the light came towards the car, almost blinding him and creating static on his car radio. He slowed down to 10 mph because he was dazzled. As the light came right over the car, it was blotted out by a huge circular object underneath it. Trevor estimates the object was about three times as big as his Morris 1000 and about thirty feet above it. But it could have been bigger and higher, or smaller and lower. The object was blue-green at one end and red at the other. He was unable to tell what it was made of and unable to see any metallic shine or rivets. As soon as the light was masked and he could see properly again, Trevor drove into Warminster as fast as he could. 'It gave me a right old breeze-up,' he admitted later. 'I was shaking all over. When I got into town I slowed down and looked back, but it had gone. There was no sound. I came along the road about the same time on Tuesday night. I went like hell, I can tell you, but I did not see anything this time.'

The lad reported his unnerving experience to his parents, then, on his father's advice, came to my house in Portway shortly before midnight. He was certainly shaken and bemused. At one point he feared that the UFO was going to strike the car, as it seemed to hover just over his bonnet and headlights.

A Red Indian prayer entreats: 'Great Spirit, help me never to judge another until I have walked in his moccasins for two weeks.' A good one for the serious investigator in any new field of research could be: 'Let not that happen which I wish, but that which is right.' After intensive study of and close contact with the UFO mystery over a long period, I am deeply conscious of the tremendous responsibility borne by all who seek to inform on such a still-futuristic subject. It is a particularly difficult task when, as is the case with Ufology, so many unknown factors have to be faced and evaluated, together with a number of incongruities and irrationalities that are an infuriating part of the UFO behaviour pattern.

Some people show not a glimmer of wonder or awe that we are hosts, willingly or unwillingly, to Outer (or Inner?), Space denizens because their eyes are blinded to reality, ears deaf to a terrific weight and wealth of first-class testimony from reputable witnesses. Personally, I feel desperately sorry for them. The shock of sudden realization, even after ample warning, may be too much for their shattered senses to stand, when it comes. To me, this disbelief is as great an enigma as UFOs themselves. How can they spurn as irrelevant and meaningless the strong evidence given by mentally healthy attestors whose characters are beyond research? A number came deliberately to disprove the modest claims made by our trio of observers to former Defence Minister Denis Healey that UFOs illuminate Warminster skies on at least half the nights of the year. Others, like a senior English lecturer, at a big American school in New Jersey, Richard George from West Orange, come in blind and rewarded faith that they would view their first unaccountable flying object in skies above a town thousands of miles away.

I am sure that whatever the eventful mission of alien spaceships and Outer Space denizens may be in our atmosphere, it cannot be wholly destructive or evil. Our UFO friends are capable of reshaping our thought patterns by media unknown and invisible to us. Providing the alteration remains

137

towards a favourable end, who are we to stop that influence penetrating deep within our subconscious selves, to the very essence and enrichment of our souls, echoing inside the chambers of minds quickened to respond?

From the emotional to the practical, here are some findings by my young collaborator, Douglas Chaundy, from Weston-Super-Mare, Somerset, in addition to his interesting discovery that constellations of the Northern sky are copied by long barrows on Salisbury Plain, dotting the terrain around Warminster. Douglas wrote to me after I had addressed a large assembly of Ufologists at Caxton Hall in London, presenting findings that go much deeper and must have great significance. Many people are aware of the celebrated White Horse cut into the chalk of hills in Britain. My Somerset friend wrote:

> The remaining White Horses of Wiltshire, together with the Uffington White Horse and Stonehenge, form a very intricate pattern that is very interesting, particularly when one considers them to have been built by people of allegedly light intelligence.
>
> The central line of the pattern is formed by the white horses at Cherhill, Alton Barnes and Pewsey, which are equidistant and directly in line with each other.
>
> On the eastern side of the line there is formed an isosceles triangle of perfect dimensions. The western side of the line forms a forty-five degrees angle of perfect dimensions, also. The two patterns are explained by Diagrams 2a and 2b. These two triangles are interesting by themselves; but are far more interesting when they are joined by the connecting line (the central line). The most interesting part of all, however, is what happens from the moment these triangles are connected to produce one complete pattern.
>
> As you can see, on the diagrams there are lines showing the outline of the triangles and the interior of them. Here is the amazing thing: if these various lines continued, and

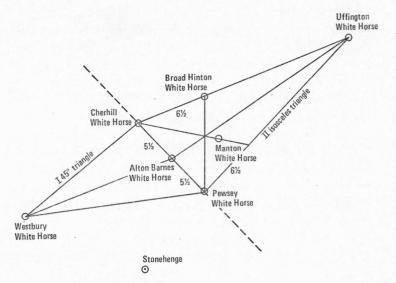

Diagram 2a (*Douglas Chaundy*)

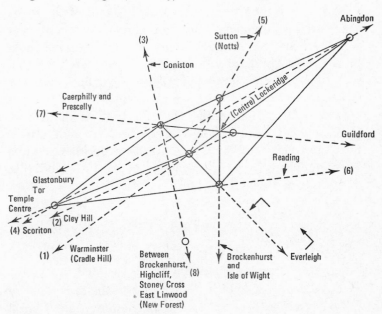

Diagram 2b (*Douglas Chaundy*)

extended from the triangles to certain localities in England, we arrive at specific points that have been visited by extra-terrestrial craft. These are not simply plain sightings, but in every case except one or two they turn out to be actual *landing* locations, starting from the white horses that are used. On Diagram 2b: 1. Warminster (Cradle Hill-Battlesbury) extended from the line that runs through Uffington and Alton Barnes. 2. Cley Hill from the line that runs through Manton and Alton Barnes. 3. Coniston from the line that runs through Stonehenge and Cherhill. 4. Scoriton from the line running through Uffington and Westbury. 5. Sutton, Nottinghamshire from the line that runs through Alton Barnes and Broad Hinton. 6. Reading from the line that runs through Westbury and Pewsey. 7. Caerphilly, South Wales from the line that runs through Manton and Cherhill. 8. New Forest from the line that runs through Cherhill and Stonehenge.

These are sighting areas that only tell part of the story. There are so many more – sightings and landings from near Guildford, at Tottenham, at Stonehenge, at Abingdon, near Oxford, etc. Apart from the sightings, there are lines that run to sundry other interesting places. Four lines go to the Glastonbury Temple of the Stars, among which is one that goes to the Tor itself, another that runs to the Zodiac centre at Butleigh. There is a line that goes to Everleigh, also to East Linwood, both mentioned in Eileen Buckle's *Scoriton Mystery*. There is also a line going to the Prescelly Hills, the origin of part of the fabulous Stonehenge. The New Forest line I have mentioned proceeds both to the location of activity between Brockenhurst and Highcliffe, and to Stoney Cross.

You will notice that the white horse at Manton, also called the Marlborough White Horse, seems to be in the wrong place. It should be, according to the remainder of the coherent pattern, at a point in the Savernake Forest. However, if this white horse *had* been in the Forest, it would have made the findings entirely wrong. By its being at the present location it is possible to arrive at locations which would otherwise have been missed. Another strange thing is the fact that the only other white horses which are in Wiltshire have been left to deteriorate; only the white horses in the pattern have survived and are being regularly looked after and tended.

The relevant and overriding question must be posed: why are the interplanetary craft employing these obvious and prominent landmarks as direction finders? And what is so special about the places where they land and where they stay for months or years on end, as is the case in the Warminster area? In your book, you express the point that it was revealed to you that there is a UFO base only a few miles from Warminster; and the peculiar triangle or pyramid shaped impressions left in rear of two leaping UFOs in a photograph mean something profound. I seriously suggest that it could be within the Triangle Pattern!

It seems that the extraterrestrials could conceivably have their beginnings *from* the triangle. I am sure that what I have found cannot be dismissed as sheer coincidence, for the simple fact that it happens so many times. I have found no fewer than seventeen lines, involving many sightings, and I am still finding them.

Essentially, Douglas is a realist, as I am; yet his view that the pyramid impressions on the UFO photograph mean something profound concurs with mine, exactly. The strong spiritual corollary to the physical properties of UFO manifestations is confirmed in various ways, as I shall attempt to explain.

Do UFOs 'home' on high frequency signals generated from apparatus buried deep underground in that area, possibly millenia ago, by forbears of our present race of UFO aeronauts; perhaps a transmitting device placed there deliberately, in shrewd foresight or calculation of a future cosmic event, scientifically computed long ago? Our team have a pretty good idea where such an instrument might be located, but realise the deadly danger resulting from foolish excavation in reaching it. By tampering with such apparatus, by attempts to remove it from its embedded power centre, the whole of the nearby barracks and an extensive section of the town itself might be blown asunder by the terrific detonation.

Another brain-teasing imponderable is: the closer the spaceship to the observer, the less chance there is of capturing a 'scoop' picture. This is amply proven by what took place – or did not, rather – at Starr Hill, between Battlesbury and Scratchbury, one night. Bob Strong took no fewer than thirty-two 'shots' of the orange-glowing and hemispherical topped spaceship that silently hovered over a shoulder of the hill in haloed glory for several minutes at about 11.56 pm. It was as bold and prominent to the eye as a lighted house window on edge, no farther than 250 yards from our vantage point near Starr barn overlooking deserted North Farm.

It showed detail in design we had never seen previously. Yet nothing developed on the plates except wavy horizontal and vertical lines indicative of scorching. Why? There could be several sound reasons, from the viewpoint of those aboard the fiery monster. The force field from the 'sitting target' may have been too overpowering at that close proximity, not sufficiently strong to harm the onlooker yet adversely affecting a camera machanism. Again, the people on board the slightly tilting craft may not have wished such revealing film (if we had been successful) to get into the wrong hands, with power-crazy cliques on Earth thereby learning too much from closeups of the exterior. Or again, they were simply demonstrating a warning of dangers from venturing too near harmful radiation.

We now have a scientist friend, employed in chemical engineering, who is working on a colour inversion principle and formula to help us decide what type of film to employ

for best results when within hailing distance of these tanta-
lizing UFOs. Colours and sounds are important criteria in
dealing with a technology and way of life that out-strips
our own in many respects, as incidents in this book will
amplify. Various entities, in a dimension beyond the infra-
red and ultra-violet, exist and are invisible to the gaze of
man at present. The blunt and inescapable truth is that
numerous facets of UFO behaviour fail to fit snugly into
conventional pockets of Earth thinking and scientific attitude.

In the county press at nearby Trowbridge, a story headlined 'Out of this World?' told us on 21 July 1978: Anybody
looking up at the skies over Trowbridge on Tuesday evening might have encountered something out of this world.
Mr Terry Berrett, of 19 The Weir, Edington, certainly did;
he saw a formation of UFOs flying low and fast across the
town. 'I was on my way from West Ashton to Trowbridge
when I saw what I first thought to be a helicopter flying
from the college side of the town across towards Hilperton.
As it got closer I saw it was a tightly bunched group of
black objects. The whole lot seemed to be surrounded by a
heat haze: the sort of haze you see when you watch the
Red Arrows go past very close.' Mr Berrett said the group of
UFOs was travelling very low and very fast. He estimated
that the things were going at 200 mph although they hardly
broke the skyline as they were travelling so low. 'I travelled
half-a-mile at 60 mph but in that time they had crossed the
whole town and disappeared behind Biss Farm.' Mr Berrett
said he had perfect eyesight and had taken a course in aircraft recognition. He said whatever he saw could not have
been a flock of birds, as they were travelling far too close and
orderly a formation, and it certainly was not a helicopter as
he had at first thought. 'It took me completely by surprise,
I still don't know what they were: I just don't know what to
make of it.'

Now let us review a faithful account by Terence Hayes with

respect to just one eventful night in the vicinity of Stonehenge. His observations cover the period 6.20 pm to 9.30 pm, 18 October 1977. Here is his report; and note the drawing he made on the spot afterwards:

There were six in the party: myself, Valerie my wife, Jane my daughter, John (my filming companion), Sylvia his wife, and Melanie their daughter. We pitched camp as shown on Diagram 3. Over the next three days of 16, 17 and 18 October, we were to witness a number of sky sightings in that locality, some of which could possibly be explained as shooting stars, satellites, meteors, etc. On the day in question, in the halflight of dusk at 6.20 pm, the first strange light appeared in the sky in the direction and relative position to our camp that my sketch illustrates. Sylvia spotted the first one, which we all watched and wondered. However, after a few moments of hovering and moving slowly to our left as we looked at the object,

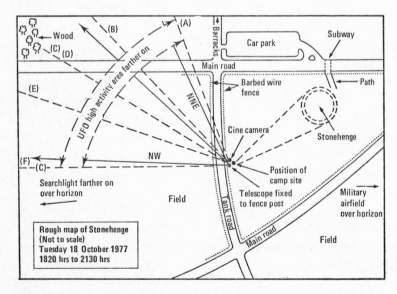

Diagram 3 (*Terence Hayes*)

144

it then simply disappeared or dematerialized before our eyes.

Minutes later, a second one hove into view, I rushed through a barbed wire fence to observe the object through a telescope that I had attached to a fence post (on the perimeter fence to Stonehenge), shown on the sketch. The weird light stayed aloft and around for at least two or three minutes. What happened next was incredible! These glowing lights came and went in the variegated formations shown in Diagram 4 (approximate positions marked on Diagram 3). Lasting a total of three hours and ten minutes, we observed many aerial perambulations.

They would abruptly stop in flight, without any deceleration; yet when moving away there was instant acceleration at sometimes extraordinary swift speed. For fast or slow rates (moving singularly as the sketch denotes), they evinced a slight, jerking motion when they changed course as seen via the telescope. They would tilt somewhat before moving away at unbelievable speed. In fact, a few I observed shot across the horizon before one could bat an eyelid! It rapidly became a mere streak of light. They could hover and change direction instantly; not in a smoothly flowing action, but rather in a swift yet jerky movement.

They would change formation, with one object at a time either dematerializing or shifting away at terrific speed (difficult for the human eye to define precisely which); then reappear to form yet another formation. All the light was contained within the sphere itself; *there was no reflected light*. Only on one occasion was any reflection discerned, which I detected thanks to the telescope; a vapour or smoke trail as shown in Diagram 4; and I saw it with the naked eye quite clearly. My wife also caught a glimpse of the trail as it faded into the then dark sky or night air.

The two girls, Jane and Melanie, had toy pocket walkie-talkies with a simple compass fixed. During the high

145

Singular movements

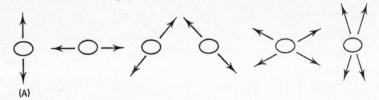

(A)

Different formations

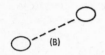

(B)

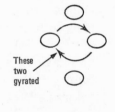

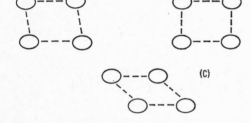

These two gyrated

(C)

(D)

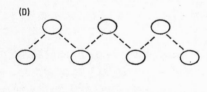

(E)

This one moved toward us and back

(F)

Object through telescope

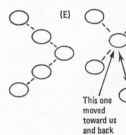

Vapour or smoke trail

Diagram 4 (*Terence Hayes*)

activity phases of the UFOs, they were unable to register magnetic North; instead the compasses went completely haywire! My friend John, who took the now-famous UFO colour movie film, had a portable television that also behaved in a most erratic manner! This seems to indicate than an electromagnetic field was either being emitted or drawn towards the unknown objects of light.

Shall we ever know, for certain? It was very difficult to assess whether we were looking at a large object far away, or a small object close to us. From time to time, just a few of the aeroforms came near to the tops of trees, hovering and moving from one side to the other at a slow speed above the treeline. One object swept very low, quite a long way from our position, and I distinctly observed through the telescope this particular one just below the treeline near the horizon. To be able to define, or even estimate the proximity in relation to the object's size, was quite a guessing game on the night in question. It would be interesting to have an expert calculate some kind of reasonable answer, if at all possible, which I doubt!

There was military activity taking place that evening, one could hear gun and shell fire, plus the engines of vehicles; you could easily discern the reflected light at a distance, over the horizon. What appeared to be a strong searchlight, was obviously aimed directly at the aeroforms or UFOs; and it was no use at all. As the beam approached within a certain proximity of the objects it faded away abysmally as if deflected by some subtle force; rather a strange phenomenon when you take into consideration that there was no reflected light from these peculiar lights in the sky. Moreover, whatever they were, the UFOs were absolutely silent during their aerial manoeuvres, even when stationary.

We were compelled to conclude that they were apparently controlled by some latent force, whether man-made, a force of nature, or a cosmic source. Maybe other phenomena unknown to man are capable of creating such

147

bizarre patterns of light, such weird shapes and substances ... from all the observations made that night it would appear that any known flying machine made by man on this planet could not possibly stand up to these extreme pressures at such terrific speed and control; even the most advanced fighter craft would disintegrate, so far as my limited knowledge of aerodynamics is aware!

The UFOs moved with such speed and precision: verily they enacted *a flying miracle before our eyes*. I stayed in my original position behind the boundary fence to Stonehenge, manning the telescope diligently until all the UFO activity ceased at approximately 9.30 pm.

Judging from the 'impossible' feats I have seen them perform on many ocasions since 1965 (altering shape, colour, speed, appearance, materializing and dematerializing at will), I honestly think the majority of UFOs are inherently of our native planet. It could be that their time clock is in advance of our own; therefore, it is a futile exercise trying to capture something from our future. We are given clues that point to these possibilities. For example, true stories like the following are becoming too commonplace and well-authenticated by witnesses of high standing to dismiss out of hand as nonsense.

It was shortly before midnight one August Saturday and thirty-six people assembled for sky-watching on Starr Hill, almost as productive for UFO sightings over the past fourteen years as the notorious focal point Cradle Hill. The sky had lightened after a cloudy evening, weather improving though sultry, and we had spotted two jerkily moving shapes high in the heavens; silvery and spherical spectres that danced an aerial jig as one jaunted south, the other moving northward, changing flight directions more than a trifle when on the verge of final blink-out. A reasonable assumption? One satellite, apparently a bit off its orbital path, our eyes misleading us; and one undisputed UFO that suddenly flared into a crimson ball high above us. Both were voiceless and aloof.

Its abrupt turning to the east, in addition to colour-changing, may have furnished a clue to the next developments, after Bob Strong and Sybil Champion alerted us on returning from a reconnoitre they had made of the isolated village on Salisbury Plain, six miles from Warminster. Shell-shattered Imber was opened to the public that August bank holiday weekend, we knew. Bob and Sybil assured us there had been an obvious landing of a fairly large and orange-glowing spacecraft several smiles from Starr. So we tagged along behind their car as it purred over the main Salisbury road and through the back road to Imber and Tilshead from the village of Heytesbury. Tension mounted within us with each passing mile; and expectations of seeing the aerial monster with gold-yellow rims were high in the minds of all.

After a few miles, warning notices erected by the Army told us not to venture off the metalled highway, for the area abounds with a dangerous dosage of unexploded missiles and minefields, tricky traps for the unwary. All approach roads and the tiny unused village – bare, desolate, remote and fit only for military rehearsals – beckoned former residents and friends that weekend, soldiers ceasing operations while former dwellers made a sad pilgrimage to the battered community currently acting as a battle practice ground for street and house-to-house warfare of guerilla type. We scarcely knew what to expect as Bob halted his car at a tank crossing not far from the rabbit-infested village and we neatly stacked vehicles line abreast on the narrow lane facing open country to our right.

I travelled with the Reverend Angus Logan of Glasgow (then senior minister for religious instruction in all Cumbernauld schools; now retired and living on Arran), his wife Christian and Malcolm Bowman, lecturer on liberal studies in Yorkshire. Dorothy Gibbs and her son and daughter from Swindon were already dismounted and being shown the approximate UFO landing site by Sybil. In the advance party were several lads from London, Middlesex and Hertfordshire, a likeable Lancastrian named Matthias Shields and his wife Stephanie, together with their four toddlers in a semi-sleeping state in a dormobile, hailing from Colne. First of the thirty-six at the scene, we eagerly scanned the area of darkness stabbed at by the forefinger of Sybil, and shirt-sleeved Bob told us it was quite big. We saw nothing in the shadowy stillness and patches of motionless undergrowth, although Sybil estimated: 'It was about three to four hundred yards away at most and as large as a house, glowing orange and shaped like a plate bottom upwards.' She expressed a view that it had moved or blacked-out while she and Bob sped to warn us at Starr.

Our avid eyes feasted on the panorama of purple-black folds of night. An owl hooted and a wild rabbit scampered through the thick bushes before us. With the aid of torches, we could clearly discern that the screen of hedgerow and bushes gave way to a wide expanse of open countryside beyond. We continued to gaze at the starlit sky, then our attention became riveted on some dancing pearly-white aeroforms in the distance that matched no car headlights we had ever seen, and which, through binoculars, were certainly an uncanny lampshade shape. In a fairly straight formation at first, they numbered three tiny discoids, almost speck-like. Never rising very high, yet one leaping up and over the others occasionally as they flowed in a smooth action from left to right, they reminded me of pearls moving on a rippling necklace through the night.

They were astonishingly pretty, petite and memorable. We did not know what they were, but everyone watching the fitful, flitting display of aerial magic felt a warm glow deep inside. Genuine UFOs all have this emotive impact on witnesses, in my experience as a journalist. In total on this Imber excursion, we counted five of them after Bob, Sybil and the main party headed for the centre of Imber Village to get a closer look at the prancing pearls of light and we had stopped at a point a few miles from the road to Tilshead.

Two silver spheres appeared on the left across mined terrain, identical in brightness and stature to the trio of forerunners. Bob later reported a lambent circular light gliding through tree growth towards Imber. Malcolm, still with the Logans and myself, logged what we were seeing as one of the light energy spheres on our left came nearer to us and dramatically turned on its edge to become a perfect circle before strangely dipping into the ground and vanishing completely. That was the outstanding clue to me, as an observer and recorder of these mystical marauders. For the single soil-scorcher and its unbelievable disappearance downward from sight set an example. Its two constant companions in triangular flight followed suit, until all three had sunk slowly and with diamond-bright deliberation, from our view. It was as if the earth had swallowed them – willing victims – piecemeal yet wholesale!

Staying there, in that idyllic spot where the sky was aflame with these remarkable light gems, although the dead ground beneath the dancing bodies was pitted and torn by shell bursts and bomb craters, until almost two o'clock, we all exulted to the sheer magic of their colourful mystery as they floated gracefully from left to right, right to left, at bewildering speed; then backtracked in the heavens at a laggardly pace, not straining the eyes of bewitched onlookers. Judging by past encounters of a visual kind, and through binoculars, I estimated their length to be no more than a yard, depth half that. They were tiny discoids that left an indelible imprint on the mind, even so; and led to my final conclusion that the majority definitely belong inherently to our planet; yet are of universal importance and significance. They have probably been on Earth since before the germination of Man; and maybe long, long ago when Ufonauts were in purely physical form as we are today, they even begat us into the wonderful, mighty Universe.

No one has yet come up with the final and unassailably accurate answer as to what UFOS are and what intelligence controls and mans these barely describable machines. If anyone on Earth possessed the answers to all the questions they pose, there would be no challenge left to resolve in life; and the answers can only come when *they* decide the time is opportune, *not* when *we* imagine it to be ... senior culture among Creation will always have what we are fond of calling 'the last word'.